NUCLEAR POWER REACTORS IN THE UNITED STATES

Energy Research & Development
Administration
December 31, 1975

UNACCEPTABLE RISK
The Nuclear Power Controversy

This is the only book that tells the complete story behind the nuclear power controversy—and tells it in language that anyone unfamiliar with nuclear technology can understand.

It is the first book to portray the extraordinary human drama behind the controversy—the gripping story of nuclear power pioneers and advocates and of men and women who have risked their jobs—and even their lives—to oppose nuclear energy.

UNACCEPTABLE RISK
The only book that tells all sides
of the nuclear energy story.

UNACCEPTABLE RISK
The Nuclear Power Controversy

by McKinley C. Olson

BANTAM BOOKS · LONDON
TORONTO · NEW YORK

UNACCEPTABLE RISK
THE NUCLEAR POWER CONTROVERSY
A Bantam Book / August 1976

ISBN 0-553-10185-4

Published simultaneously in the United States and Canada

Bantam Books are published by Bantam Books, Inc. Its trade-
mark, consisting of the words "Bantam Books" and the por-
trayal of a bantam, is registered in the United States Patent
Office and in other countries. Marca Registrada. Bantam
Books, Inc., 666 Fifth Avenue, New York, New York 10019.

To Mary, Miles, Caleb, and Josiah

Contents

Introduction

"A serious threat to the future of all life on this planet." These were the words used by a highly respected nuclear technician to sum up his many years of experience working with nuclear fission power plants.

He was one of three engineers who threw away their salaries, careers, and job security in February 1976 to join the growing crusade in this country against nuclear power. Together, the three men had over 50 years of experience with General Electric Company's nuclear reactor division. They held important positions in management. And all three had growing families to support. But they were driven to resign because they believe nuclear plants—and most everything that goes with them—are intolerably dangerous.

One of the men said, "We can never build them safe enough." Another called nuclear fission power "just too big a risk"—a "technological monster" whose "potential disasters are so vast, it is beyond my comprehension." A major nuclear accident, he warned, could "kill tens of thousands of people in hundreds of square miles."

And the three had other concerns to tick off about nuclear power: the dire necessity of containing and safeguarding radioactive nuclear waste for hundreds of thousands of years; the largely unexplored relationship between radioactivity, disease, and genetic defects; the dangers of human error, accidents, sabotage, blackmail, and theft throughout the complicated nuclear fuel cycle; and the certainty that the spread of nuclear-power reactors would accelerate the proliferation of nuclear weapons around the globe.

A week after they resigned, a nuclear engineer who supervised the construction of nuclear plants for the federal government was on nationwide television saying that nuclear plants were poorly designed and built; and that he was going to hand in his notice because nuclear power is inherently unsafe.

Within days of their resignations, the four uneasy nuclear engineers were calling for a national moratorium on the construction of nuclear plants, a phaseout of those in operation, and programs to conserve energy and employ other sources of power such as solar energy and the antipollution use of coal.

The powerful multibillion-dollar nuclear establishment, reacting angrily to the widespread publicity engendered by these resignations, claimed the charges were unsubstantiated, and insisted that nuclear power was safe and essential to the economic well-being of the nation and the world at large.

But the chairman of the powerful Joint Committee on Atomic Energy, one of the cornerstones of America's nuclear edifice, replied that "the public is aroused and we have to satisfy the public that nuclear power is safe."

This may be impossible, because the public *is* aroused. Citizen critics are pushing antinuclear legislation in some 20 states and putting the pressure on Congress to abandon its long-standing support of nuclear power. And newspapers, magazines, radio, and television are issuing a steady flood of reports about the nuclear controversy.

Ominous cracks appear in the vital cooling pipes of a reactor unit and nuclear plants around the country are forced to shut down for inspections. The flame from a small candle starts a fire in the South that sweeps through layers of vaunted safety systems backed by systems piled on top of systems to dangerously cripple the world's largest nuclear plant.

In 1976 there were 56 nuclear plants licensed to operate and some 60 under construction in the United

States. They supplied us with about 8 percent of our electricity. If the nuclear proponents have their way, there will be about 1000 large-scale nuclear plants in this country by the year 2000, providing us with at least 30–60 percent of our electricity. By then, it might be too late to turn back.

It's coming to a head. For you and me and our children and generations to come.

I
The Credibility Gap

People from all walks of life are getting caught up in the nuclear controversy. In California, Edward G. Stotsenberg says, "The opposition to nuclear power seems to be growing out here." He's a trim, conservative certified public accountant, horse trainer, and budding sculptor who has a house on the beach at Malibu where he lives with his attractive and hospitable wife Dorothy.

"You know how people talk—about how California's going to slide into the ocean when the next big earthquake comes along. Well, it's serious," Stotsenberg said in 1975. "California's riddled with latent faults. We get a big jolt out here at least every four or five years. Some of them are really serious. And yet the Los Angeles Department of Water and Power wanted to build a nuclear-power plant right on top of an old fault in the west end of the county at a place called Corral Canyon."

The 60-year-old Stotsenberg pointed north along the shore toward the canyon. "That's 5 miles from our house," he said. "I know because I jog up there and back all the time.

"The fight over the plant began in the early sixties and dragged on and on. Some of us who live around here were worried about thermal pollution from a nuclear plant; or that radioactivity could escape. And earthquakes—that an earthquake could cause a major nuclear accident. Then some people in the community hired a geologist, who proved that the plant would sit right on top of this old geologic fault. So they finally dropped their plans to build a plant at Corral Canyon.

"But it was quite an eye-opening battle while it lasted, I'm telling you. We learned a lot about nuclear plants. They're just too dangerous to suit us. And the people who are for 'em just don't seem to know what they're talking about half of the time."

This fear of earthquakes and nuclear power in our most populated state was joined back in 1958, when Pacific Gas and Electric Company in northern California, the second largest public utility in the country, began purchasing land to build a nuclear fission power plant on Bodega Head, a beautiful, isolated peninsula on the coast about 60 miles north of San Francisco.

The site of this proposed 340-megawatt (340,000-kilowatt) reactor power plant was only a few thousand feet west of the edge of the famous San Andreas fault, which caused the great San Francisco earthquake of 1906 and continues to be the most active source of quakes in the country. Many people were convinced that an earthquake could trigger a catastrophic release of radioactivity if the plant were built; and opposition to the reactor unit began to grow.

The utility had committed more than $4 million in planning and construction costs to the project. It kept insisting that a nuclear-power plant could be built to ride out any earthquake without a rupture, even though the crust of the earth had shifted more than 20 feet in some places during the great quake of 1906, when some sides of the fault wound up 2–3 feet higher in spots when the trembling earth finally came to rest.

The Bodega Head controversy began to spread around the state. Thousands of Californians, and many people elsewhere, were treated to a flow of reports on the dangers of nuclear-power plants and their related facilities that were staggering in their implications. And the fact that the nuclear establishment seemed so determined to build a reactor power unit in such a high-risk earthquake zone made many ordinary citizens begin to question the credibility of everything connected with the country's nuclear-power program, no matter what or where.

Then the U.S. Atomic Energy Commission, which had initially approved the project even though the site was too close to the San Andreas fault to comply with the AEC's own regulations on power plants and earthquakes, began to reverse itself; and doubts were expressed by the agency in 1964 about the wisdom of building a nuclear plant on the peninsula.

Finally, on October 30, 1964, after six years of bitter struggle and controversy, Pacific Gas and Electric reluctantly announced it was going to give up on Bodega Head. But the suspicions and the doubts that grew up around the prolonged controversy have refused to die. They keep coming back to haunt the credibility of the nuclear establishment whenever it begins to talk about building another nuclear plant, no matter how and no matter where.

II
Legislative Warfare

In the spring of 1975 a woman with two young children at her side walked out of a Sacramento supermarket pushing a cart full of groceries, and saw she would have to walk past some neatly dressed volunteers who were standing behind a portable table decked with literature and a sign that said, "People for Proof—the California Nuclear Initiative." When she paused, one of the volunteers asked her if she was registered to vote in the state. The woman said she was. "Then would you mind signing this," the volunteer said, handing her a petition.

"Isn't this the campaign to outlaw nuclear-power plants?" she said.

"No," she was assured. "It's not against anything. It's just a petition for safe nuclear power. All we're asking for is demonstrations—proof, that nuclear power is as safe as its advocates say it is."

"Well, that certainly sounds reasonable to me," she said, adding her signature to the list. "And I'm sure my husband will feel the same way."

Scenes like this were repeated thousands of times or more around the state in the winter and spring of 1975, according to Dwight W. Cocke, a young bearded ex-actor turned full-time nuclear critic from San Francisco, who wound up as the northern California director of the Californians for Nuclear Safeguards—otherwise known as People for Proof.

Around the same time the woman was signing the petition in Sacramento, a colorful, controversial political activist named Edwin A. Koupal, Jr., whose trade-marks are cunning, a brash tongue, and long white bushy sideburns, was in Los Angeles, addressing some archconservatives who had invited Koupal to give them his views on the nuclear-power controversy in California.

"We believe that nuclear utilities should be as safe as any other businesses in our communities," Koupal was saying. "And the fact that the federal government has a special insurance program to subsidize nuclear-power plants and limit their liability means one of two things—either nuclear-power reactors are too danger-ous to insure, or the federal government has singled out this special industry in order to promote socialized in-surance."

Koupal grinned. "They bought it," he told me. "Those Birchers lined right up to sign the nuclear initia-tive. You just have to know how to reach people, to put the issues to them in terms they can grasp and appreciate."

Dwight Cocke and Ed Koupal were among the strategists I met from the California nuclear initiative, which was able to attract and direct thousands of vol-unteers, from a host of cooperating organizations, who fanned out across the state with their petitions. They were taking advantage of the fact that California was one of 22 states in the union with the initiative process, which offers people the opportunity to write their own laws and put them to a direct test of the voters if they collect enough names to qualify for the ballot.

On May 6, 1975, California's Secretary of State March Fong Eu announced that the nuclear safeguards initiative, which had gathered more than 500,000 sig-natures, had qualified for the June 1976 statewide bal-lot. Approving the measure meant a moratorium could be placed on the development of nuclear power in California.

Under the proposed Nuclear Safeguards Act, which would exempt small-scale medical and experimental re-actors from its provisions, the utilities would have to provide enough proof to convince two-thirds of the California legislature that nuclear-power plants and their related facilities and functions were safe enough to operate. And the act specifically called for the following:

• Demonstrations that emergency safety systems at nuclear plants would actually work when needed.

• Assurances that every step of the nuclear fission process could be protected from the dangers of accidents, disease, sabotage, and theft.

• Guarantees that the radioactive waste produced by reactor power plants could be safely stored or permanently disposed of.

• Enough liability insurance to fully cover people in the state in the event of a nuclear accident.

All these add up to what much of the controversy over nuclear fission power is all about in the United States. And it is doubtful that the nuclear establishment would be able to comply with the standards required by the act. These are some of the reasons why.

All the model-scale tests of emergency cooling systems for nuclear reactor plants have failed. And the American Nuclear Society claims that it would cost about $250 million (about one-fourth the cost of a large new nuclear plant in 1975) for a full-scale test— "impractical" and "prohibitively expensive," in the words of the society—because a large part of the plant cooling system would have to be destroyed for each test, which might damage the reactor too. So no one is positive that these backup safety systems would work in a real life-and-death emergency; and the available evidence suggests they might not.

Trying to protect the public from exposure to radioactive materials through all the complicated links in the nuclear fuel chain would be a monumental and perhaps impossible task. Many critics are especially concerned about radioactive plutonium 239, a man-made by-product of the fission reactor process. Our operating nuclear plants created around 12,000 pounds of plutonium in this country alone in 1975. And this could reach an annual figure of around 400,000 pounds from the type of reactors now operating in the United States by the year 2000, according to government estimates.

Plutonium is the most dangerous substance handled in quantity by man. A speck of it is enough to cause cancer. It must also be guarded from theft, because less than 20 pounds of plutonium 239 is enough to make a homemade atom bomb.

No one seems to know what we're going to do with the radioactive waste from nuclear plants. Temporary storage facilities—most of them large underground tanks cooled by water—have leaked; and the experts haven't been able to figure out a way to permanently dispose of this waste. It will have to be perfectly contained, too. Some of these deadly radioactive poisons can cause cancer, leukemia, and heart disease hundreds of years from today. Plutonium 239 will be able to do this a hundred thousand years and more from tomorrow.

The most controversial part of California's nuclear initiative pertains to liability insurance. A federal law, the Price-Anderson Act, limits the total liability of a single nuclear power plant accident anywhere in the nation to $560 million, of which a utility would only be liable for $125 million in 1975 because taxpayers' money would cover the rest. A major nuclear accident could cause billions of dollars in damages. This liability limit means the public would be able to collect only pennies on the dollar. The law prohibits the public from suing to recover for damages. People could lose their homes, cars, businesses—to say nothing of their very lives and health.

The California initiative would lift this shield. People could sue to recover for damages no matter how high the amount. The utilities would have a year from the day the law went into effect to obtain coverage for their plants; if they did not, the state would ban the construction of more nuclear plants; and the operating capacity of those in existence would be lowered after one year to 60 percent for five years, and then go down another 10 percent each year thereafter.

Everyone agrees that losing this federal liability ceiling, with its protection against claims in excess of $560 million, could destroy nuclear power in California, because no private insurance pool would provide enough liability for nuclear plants, although the nuclear establishment claims their plants are safe. The risks are simply too great, and the losses would be ruinous, which is why the federal government has limited nuclear liability and subsidized it. Without this protection, atomic energy would never have been able to develop in the United States.

Ed Koupal raves about the "beauties" of the initiative process. In 1969, when he was training car salesmen, Koupal discovered the initiative process on the California statute books, which he felt could be dusted off and used for public interest issues. Enthralled by its possibilities, Ed Koupal and Joyce, his lanky wife and coworker, formed the People's Lobby and promptly set out to reform as much of their world as they could.

"One of the great things about the initiative," Koupal explained, "is that ordinary citizens can go directly to the people with the great issues of the day, like nuclear power—bypassing the politicians, powerful lobbies, special-interest groups, and state legislatures. Another is that it avoids the frustrations of lobbying and protesting. You know right away—at the signature-gathering stage, or at the polls—if you've won or lost. If you lose, you regroup, learn from your mistakes, and begin all over again until you finally win. And when a people's initiative wins at the ballot box, it becomes law. It has to be obeyed, like it or not." Koupal added that the antinuclear coalition would not stop trying to pass a nuclear initiative until it became law in California, and in other states as well which have the initiative process.

In 1972, the People's Lobby included a five-year moratorium on the construction of nuclear-power plants as part of a Clean Environment Act for California. The initiative qualified for the ballot, but was defeated at the polls by a 2–1 margin.

"Afterward we ran a poll and discovered that the

word 'moratorium' confused people," Koupal said. "And that it had negative connotations. Some people thought it had something to do with the war in Vietnam; others were afraid it meant the utilities would close down their plants and shut off the lights. And when people are confused or afraid, they either vote 'no' in favor of the status quo, or don't vote at all."

In another area, Common Cause and the People's Lobby, with its 15,000 members and working affiliations with other groups in the state, bounced back in 1974 with a political campaign initiative that gathered a half-million signatures to qualify for the ballot. This initiative was bitterly opposed by big business and big labor, which particularly objected to the sharp limit the measure would impose on campaign spending. Nevertheless, the initiative went on to win at the polls by a 3–1 margin to give California one of the toughest political reform laws in the United States.

Buoyed by the successful campaign for political reform in California, the People's Lobby began a drive to amend the U.S. Constitution to permit citizens to petition for new laws, repeal existing statutes, and remove public officials from office—which won the endorsement and support of Ralph Nader and his Citizen Action groups because it could put direct power into the hands of the people.

In the early spring of 1974 there was another attempt to revive the antinuclear movement in California "that got nowhere," according to energetic Jim Harding, the young, knowledgeable energy projects director for the San Francisco-based Friends of the Earth environmental group. "We spent three months running around the state for a piddling 20,000 signatures. We called a statewide meeting of our supporters, decided to throw in the towel and start all over again."

"We spent too much time trying to convince the media to take a stand against nuclear power instead of going right out after signatures," Harding explained. "Also, we were disorganized. Some of us wanted a moratorium on nuclear power; others were afraid that wouldn't attract enough support."

"You just can't sell a negativism like 'moratorium,'" Ed Koupal kept stressing, adding, "promotion and sales—that's what politics is all about."

The antinuclear activists formed People for Proof; lawyer Richard Spohn, a former Ralph Nader associate, drew up the legal terms of the initiative; and three months later they had gathered more than 300,000 signatures. The California nuclear safeguards initiative, coordinated by Californians for Nuclear Safeguards, headed by lawyer David E. Pesonen, a long-time antinuclear activist, attracted the support of groups such as the California Citizen Action Group, the Friends Legislative Committee, Another Mother for Peace, Project Survival, Zero Population Growth, and environmental organizations such as the Sierra Club and Friends of the Earth; celebrities like Robert Redford, Steve Allen, and Jack Lemmon; scientists like Nobel laureate biologist Dr. Linus Pauling and high-energy physicist Dr. Henry W. Kendall from Massachusetts Institute of Technology and the Union of Concerned Scientists, biologist Paul Ehrlich from Stanford, and former Atomic Energy Commission scientist Dr. John W. Gofman from the Committee for Nuclear Responsibility.

Big business and labor were bitterly opposed to this initiative too, and tried to capitalize on their allegation that driving nuclear power from the state would drive away jobs. Robert H. Finch, former lieutenant governor of California and former U.S. secretary of health, education, and welfare, said the nuclear initiative would "deny Californians the use of nuclear energy at a time when unemployment and other hardships could be greatly alleviated by this low-cost, clean, and unlimited source of energy . . ." Sigmund Arywitz, executive secretary of the Los Angeles County AFL-CIO Federation of Labor in 1975, called the initiative "a proposition that would in effect eliminate any hope of developing nuclear power in the state. . . . The strength of our economy—jobs and paychecks—is very much at stake." And ex-California governor Edmund G. (Pat) Brown, chairman of the California Council for Environmental and Economic Balance, which was combating the initiative, charged it was illegal "because the federal government has sole authority to regulate nuclear power, under present interpretations of the Constitution. . . . As a result, the initiative, if it passes, will most likely be overturned in the courts—but not before disastrous delay."

The courts have said that federal nuclear regulations preempt state statutes on nuclear safety. In Minnesota, for example, the courts ruled against a law passed by the state because it was stricter than the federal standards governing the amount of radioactivity that could be routinely released from nuclear-power plants. But the California initiative hinged on the right of the state to rule on "rational land use planning" to demand that reactor plants conform to the same standards expected of other industries and developments in the state.

Other states have developed their own strategies. Vermont passed a law that nuclear plants cannot be built without the approval of the state legislature. Wisconsin now claims the right to weigh the safety of nuclear-power plants in relation to the reliability and cost of electricity in the Dairy state. And voters in New Jersey turned down plans to build offshore reactors.

And most of the critics say that people who claim that jobs and the future of the economy depend on nuclear power are trying to peddle a bogus issue. The critics point out that nuclear power was only supplying the nation with 2 percent of its total energy (and 8 percent of its electricity) in 1975; and that nuclear power does not represent an immediate source of electricity in that it takes around ten years (including planning) to build and operate an atomic plant.

But more important, according to the nuclear opponents, is their argument that there are any number of safe, viable alternatives to nuclear power—especially energy efficiency programs and conservation, which could assure us plenty of energy for jobs and a high standard of living. Here they add that nuclear power would probably be abandoned in the event of a major nuclear accident—which in turn would have a disastrous effect on the economy if we had reached the stage of overdependence upon nuclear power by that time.

"We're especially lucky in California because we have the sun most of the time to give us solar energy," I was told in Los Angeles by Jeanine Hull, a busy young feminist who was the southern California director of People for Proof in 1975. "And we're filling up our beautiful canyons here with garbage which could be a terrific energy resource. A lot of people are going to come our way," Ms. Hull predicted, "when they realize they can cut the umbilical cord from the electric companies, when they see we can convert garbage into energy and draw direct power from sunshine."

"We actually have two campaigns," Ms. Hull explained. "One to alert people to the dangers of nuclear power; the other to convince them that alternative sources of energy are available. And now that the initiative has been qualified for the ballot, the antinuclear movement has a real flair and the flush of success."

Dwight Cocke, up in San Francisco, told me that "most of the signatures for the initiative were collected in the north. But the final power and the votes are in the south." Ed Koupal agreed with Cocke. "Forty-six percent of the registered voters in California live in Los Angeles County," Koupal said. "That's why the People's Lobby is here. It's a great place to be if you're really trying to change the system—to make it more responsive to what citizens actually want."

Koupal went on to discuss the importance of California in the context of the national antinuclear movement. "One out of every ten Americans lives in the state. It has a land area bigger than Japan's. We produce 10 percent of the gross national product. We have the sixth largest government budget in the world. It has every political persuasion in the country—especially around Los Angeles. It's obvious, then, that if you can get an antinuclear movement going here, it has a good chance of going anywhere."

Koupal was working especially hard, when I saw him in 1975, to spread the antinuclear gospel throughout the West—and in the East too. He was doing this as the director of an organization called the Western Bloc. It was planting seeds for antinuclear campaigns in Washington, Oregon, Idaho, Nevada, Arizona, Nebraska, Oklahoma, Utah, Colorado, Wyoming, Montana, North and South Dakota, Kansas, Missouri, Arkansas, Michigan, Iowa, Maine, and Massachusetts. Seventeen of these states have the initiative on their statute books. Oregon qualified a Nuclear Safeguards Act for the November 1976 statewide ballot in record time. All told, some 20 states, with Koupal's help, were pushing antinuclear legislation in 1975 and 1976.

"Here, take a look at this map of the United States," Koupal told me in the bustling headquarters of the People's Lobby, with its modern print shop to take on commercial jobs that help fund the group's political action activities and churn out a steady stream of literature and propaganda for its own and related causes. "See," he said. "Two-thirds of the land area in the United States is west of the Mississippi. That's why the Western Bloc is so important. When such a large part of the country begins to speak out against nuclear power, it's bound to have an important political effect. Candidates will have to take positions on the nuclear issue. And the national press will not be able to mistake the importance of such a uniform people's movement, united against nuclear power."

Koupal sounded very confident. I guess that's his real trademark.

III
Nuclear Plants

I was standing at a place called Peach Bottom, looking up at one of the biggest nuclear-power plants in the world. These silent plants are huge. Stupendous. Here or anywhere. It literally takes your breath away to look at them. And some people find the quiet that surrounds them eerie.

This one sits on the west bank of the Susquehanna River, in the southeast corner of Pennsylvania. In essence, nuclear plants, like the one here at Peach Bottom, are much alike. They all harbor enough radiation, after they've been operating a while, to kill and cripple thousands of people in the event of an accident. And their purpose, of course, is to boil water into steam to produce electricity.

Let's take a look at the Peach Bottom plant. We can see cooling towers and ponds, administration buildings, small units that house auxiliary emergency power systems and pumps, and a long smooth rectangular windowless building that contains the twin 1065-megawatt boiling-water reactors, each with its own control room, massive turbine, and generator under this one long roof. This containment building makes the railroad cars standing next to it look like Matchbox toys.

The reactors, which were made by General Electric Company, are below the ground inside the great hall. They too are enormous. The stainless steel vessels are 73 feet high and 21 feet in diameter. The steel walls are more than 6 inches thick. Each reactor weighs around 700 tons.

Philadelphia Electric Company owns the largest percentage of this $900 million nuclear complex, which the utility operates in the southern tip of York County, just above the Pennsylvania-Maryland line, 38 miles up the Susquehanna from Baltimore and 63 miles southwest of the Philadelphia area, where more than a million of the company's customers live. These are the people who get the electricity from Peach Bottom; those who live around the plant do not.

The first time I visited Peach Bottom was in 1959. I was a reporter then for a famous little morning newspaper called the *Gazette and Daily,* published in the city of York, about 35 miles northwest of Peach Bottom, and I had been assigned to cover Philadelphia Electric's first venture into nuclear power with the construction of an experimental high-temperature gascooled 40-megawatt reactor plant which began generating in 1967 and was shut down in 1974 after the utility said it was too small to be of commercial value.

The scientists tell us that each one of these 1065-megawatt reactors at Peach Bottom will produce as much radioactivity in an operating year as we would get from the fallout of 1000 Hiroshima-sized atom bombs. Dr. Henry W. Kendall from MIT in Cambridge, one of the leading critics of nuclear power in the United States, says, "The uncontrolled release of even 5 or 10 percent of this inventory could bring instantaneous death to persons up to 60–100 miles from a large fission-power reactor. Persons hundreds of miles distant could suffer radiation sickness, genetic damage, and increased incidence of many diseases including cancer."

To put this site into perspective, 8000 people live within a 5-mile radius of Peach Bottom, 5 million within 50 miles; and Washington, D.C., for example, is only 60 air miles away from these Peach Bottom reactors. Thirty percent of the nation's 200 million people live within a 250-mile radius of this nuclear-power complex. But all large operating reactor plants are essentially the same, no matter where. They all contain a tremendous amount of radioactivity, and they are all relatively close to areas where many people live. In the event of a major nuclear-power plant accident, a lot would depend on climatic conditions—especially on which way the wind was blowing.

We know that people who have been exposed to large doses of high-level radiation poisoning have suffered exceedingly painful deaths. The symptoms often include nausea and vomiting. The bowels turn to water. Victims suffer fever, prostration, stupor, and hysteria. Burns reduce the number of white cells in the body, lowering the defense against infection. The hair falls out. Then delirium, and finally death.

Low-level radiation takes much longer to take its toll, but the consequences can also be extremely agonizing. Most of the symptoms don't appear for 15 years. Sometimes it takes over 30 years for tumors to develop. This low-level radiation can cause cancer, leukemia (within five years from the time of exposure), and diseases of the heart. Victims have lost their fertility, aged prematurely, developed cataracts, and suffered anemia.

There is no chance, however, for a conventional nuclear-power plant to blow up like an atom bomb. The bomb uses uranium that has been enriched to 97 percent pure in order to make it explode, whereas most power plants use uranium fuel that has only been enriched to 3–5 percent, which is far from enough to make it explode.

There are other differences, and similarities, between an atom bomb and a nuclear-power plant, both of which fission—or split—uranium atoms in a proliferating chain reaction to create an astronomical amount of high-temperature heat energy. The bomb is designed to fission in less than a second. Theoretically, it would take five years or more for a reactor to consume the uranium in its fuel core. Also, the fissionable material is arranged quite differently. The heart of a bomb is shaped in the form of a compact sphere—which gives it the best chance of going off. Triggering the bomb with an explosive jams the atoms together under high pressure so that they unite in what the physicists call a critical mass to blow apart. It takes about 37 pounds of very highly enriched uranium 235, or 17 pounds of reactor-grade plutonium 239, to form a critical mass that can trigger an explosive nuclear chain reaction.

In a reactor, millions of small uranium dioxide fuel pellets are arranged vertically in long slender rods that are separated from one another to permit thousands of gallons of water—or sometimes gas—to flow between the fuel rods to keep them from overheating, fusing together, and melting, which could trigger a frightening sequence of events that could release a catastrophic amount of radioactivity into the environment.

The big difference, then, between the bomb and our contemporary reactor plants is that the bomb is dangerous when it becomes critical and the reactor if it begins to melt. And there have been a number of serious mishaps at nuclear plants to indicate that this dread meltdown can happen.

In 1966, for example, the Enrico Fermi experimental fast-breeder reactor (which was designed to produce electricity and create more fissionable fuel material than it consumed) had to be shut down when its fuel core partly melted. Part of the plant was flooded with high-level radiation. The people in charge of the Fermi reactor, and many others, were terrified: they were afraid the reactor would "run away." Detroit was only 30 miles away from the plant, which is on the shores of Lake Erie. A University of Michigan study calculated that 133,000 people could have been killed if the reactor had gone totally out of control. Again, this would have depended upon the direction the wind was blowing. Investigators later discovered that part of the fuel core began to melt because the flow of coolant had been blocked. The Fermi plant was closed for good in 1972.

Nevertheless, the nuclear utilities keep insisting they wouldn't be operating fission reactor plants unless they were perfectly safe. The key word is "perfectly," because a nuclear plant has to be virtually immune to human and technological failings in order to protect the public health and safety. Most of the people who are opposed to nuclear power in this country believe it is impossible, in the long run, to attain and maintain these theoretical standards of perfectibility.

There have been three major studies thus far, all commissioned by the AEC, delving into the hypothetical consequences of a major nuclear-plant accident. The first, presented to Congress in 1957 by the federal Brookhaven National Laboratory on Long Island, concluded that a runaway 200-megawatt reactor 30 miles from a city could kill 3400 people within a 15-mile radius of the plant, injure 43,000 within 45 miles, contaminate an area the size of Maryland (taking it out of use for hundreds of years), and cause up to $7 billion in property damages.

In 1964, the AEC asked the Brookhaven National Laboratory to update its first report on the basis of larger reactors that had been developed since then. The second study reported in 1965 that a major atomic-plant accident could kill 45,000 people, injure 100,000, blanket a state the size of Pennsylvania with radioactive poisons, and cause $17 billion in property damage.

This was the report the AEC withheld from the public until 1973, when a freedom-of-information suit was filed that forced the government to release the study. The critics say the nuclear establishment—the AEC, the Joint Committee, and the nuclear industry—was reluctant to disclose the report in 1965 because it was afraid that public reaction to its terrifying findings might have killed the development of nuclear power in this country.

Before it was abolished, the AEC commissioned a third report, this time from a study group headed by Dr. Norman C. Rasmussen, a nuclear physicist at MIT. The Brookhaven reports tried to gauge the worst that could happen. The Rasmussen study avoided this, concentrating on the probability of a nuclear accident occurring, and the likely consequences if one did. It took into account the fact that there were 15–20 million people living within 20 miles of the nuclear-power plants in the United States. Two large reactor plants were used in this theoretical study. One of them was a Peach Bottom reactor.

On August 20, 1974, the first public draft of the 3300-page Rasmussen report was released. One part said, "The likelihood of being killed in any one year in a reactor accident is one chance in 300 million, and the likelihood of being injured in any one year in a reactor accident is one chance in 150 million. . . . For accidents involving 1000 or more fatalities, the number is one in 1 million or once in a million years . . . just the probability that a meteor would strike a U.S. population center and cause 1000 fatalities."

The casualty figures presented by the Rasmussen report said 3400 people might die from the "worst" nuclear-plant accident that could happen, 5600 more could suffer acute radiation injuries, another 300 genetic defects, and property damage might be in the neighborhood of $6 billion.

The nuclear establishment was pleased by the Rasmussen draft report because its findings were much less horrendous than the updated Brookhaven report of 1965. But the critics were not. Some pointed out that the study did not consider the probabilities or consequences of sabotage. Others said the Rasmussen report neglected to probe the possible cumulative toll from cancer and leukemia that would be triggered by radioactive cesium 137. The American Physical Society, whose special task force identified this omission, said an "average" (not the worst) accident would cause 10,000–30,000 fatal cases of cancer, 3000–20,000 genetic defects, and 22,000–350,000 cases of thyroid irradiation in children.

The federal Environmental Protection Agency reviewed the Rasmussen study and concluded that casualties could be ten times as great—69,000 people could be killed rather than 6900. And a critique prepared by the Union of Concerned Scientists and the Sierra Club, including experts from MIT and the Rand Corporation, estimated that the chances of a major accident occurring were 100 times as probable as the Rasmussen report claimed they were. David D. Comey, a well-informed critic from the Chicago-based Business and Professional People for the Public Interest, put it this way: "Based on the number of reactors expected to be in operation by 1990—700 of them— if nuclear industry projections are fulfilled, we would be having one such accident every six years. In any case, whether it's 6900 or 69,000 people killed, I don't think the American public would tolerate that as the price for generating electricity."

The final version of the Rasmussen report, formally known as the Reactor Safety Study, was released by the government on October 30, 1975. Some of the figures in the final version were larger than those presented in the draft. The worst accident considered in the final report could cause 3300 "early" (as opposed to long-term) deaths, 45,000 cases of cancer, 5100 genetic defects, and $14 billion in property damage. And the Rasmussen report concludes that the "worst" accident has a chance of occurring about one in a billion per reactor per year.

At the root of the controversy is the nuclear fission power reactor itself, which, essentially, is a giant boiler and pressure cooker. Almost half of the reactors in the United States are boiling-water models, most of them made by General Electric, which produce steam at temperatures of around 550° F under pressures of up to 1000 pounds per square inch. Ordinary water —which the scientists call light water—is pumped into the bottom of the boiling-water reactor vessel, rises up, around and through the hot fissioning radioactive fuel core to cool it, and carries its heat to the top of the reactor, where the water boils into steam which travels directly to a turbine.

The pressurized-water reactor is the other type of light-water unit in common use. Most such reactors have been made by the Westinghouse Corporation. Tremendous pressures are exerted to keep the water from boiling into steam inside the reactor vessel in order to increase the heat efficiency of the water, which flows through a heat exchanger that turns water in a secondary system into steam to spin a turbine generator.

Another kind of reactor uses gas instead of water to cool the fuel core and transfer its heat to boil water into steam. In principle, all nuclear-power plants are the same: they create and produce energy—from heat to mechanical to electrical.

The fuel core is the heart of every reactor. Basically, the conventional American models of today are all the same. The two 1065-megawatt reactors at Peach Bottom are boiling-water units. Each core here contains over 10 million uranium dioxide fuel pellets about ½ inch in length and diameter. Each pellet has a theoretical energy equivalent of about 3 tons of coal, 12 barrels of oil, or 157 gallons of regular gasoline. These cylindrical pellets are encased in pencil-thin metal rods—often called pins—about 12 feet long, which have been assembled into 764 bundles. All told, each fuel core at Peach Bottom contains about 160 tons of uranium.

Only a tiny amount—less than 1 percent—of the uranium found in nature can be fissioned. This is uranium 235. The rest is uranium 238, which the scientists call fertile because it can produce the fissionable metal called plutonium. And most of the atoms inside the fuel pellets are uranium 238.

Everything is made of atoms. There are billions of billions of them in a single breath of air. The center of the atom is the nucleus, which contains particles that carry a positive charge called protons, and neutrons which have no charge. Particles that carry a negative charge, called electrons, go orbiting around the nucleus. Each stable atom has the same number of electrons and protons. This binds them together in the atom.

The nuclear chain reaction inside the reactor begins when a neutron is released with enough force to shatter the energy that binds a uranium 235 atom together. When the atom splits, it releases two or three of its neutrons, which go shooting out at moderated speeds of about 10,000 miles an hour. These go smashing into other atoms, shattering them into lighter radioactive elements to produce a tremendous amount of heat while releasing more and more neutrons to keep the nuclear chain reaction going.

The uranium 238 atoms have a tendency to absorb fast-moving neutrons, which would spoil the chain reaction. This is where water or gas comes in; it not only cools the fuel core and transfers its heat, but keeps the neutrons from moving too fast—just slow enough at 10,000 miles an hour to keep them bouncing off most of the uranium 238 atoms they encounter, but fast enough to split a uranium 235 atom when they happen to meet up with one.

Each reactor is equipped with control rods that are used to start, regulate, and stop the nuclear chain reaction. They contain a material—such as boron—that can absorb neutrons. There are 185 control rods inserted between the fuel assemblies in each reactor core at Peach Bottom. Neutrons are released to begin the chain reaction when the control rods are withdrawn from the core. When the reactor is running the rods are positioned to keep the fission process going at an even, efficient rate. To stop the reaction, the control rods are pushed all the way in to absorb the neutrons and stop the nuclear process.

Although uranium is radioactive, the bundles of fuel rods are comparatively safe to handle until they are placed in a reactor and undergo fission. Then the centers of the uranium pellets reach temperatures of around 4000°. And when the uranium 235 atoms split, they form a large number of highly radioactive elements —among them strontium 90, cesium 137, and iodine 131—while each uranium 238 atom that is able to absorb a neutron is converted to plutonium 239.

All told, a large nuclear reactor plant produces about 2000 pounds of these deadly radioactive poisons in an operating year. This material is too hot and too dangerous to be touched by human hands. It must be cooled at all times and handled by remote control. All these radioactive poisons must be contained—inside the reactor and after they leave it. Trying to prevent them from escaping is what most of the nuclear nightmares are all about.

The nuclear utilities claim they can bottle up enough of this radioactivity to produce safe electricity. And they point to all the containment barriers at their atomic plants. The fuel pellets are ceramic to imprison most of the radioactive gases and other fission products. The pellets are sealed in long metal tubes—usually made from a zirconium alloy—to contain the radioactive discharges from the pellets. Some of these fuel rods, however, will sprout pinhole leaks, through which the poisonous elements will escape into the coolant. And from there a small percentage will be discharged into the environment.

The fuel core, in turn, is surrounded by the reactor vessel, whose stainless steel walls are 6–8 inches thick. This sits in the middle of another shield made of steel and reinforced concrete about 10 feet thick to protect operators and equipment at the plant from radiation. And all this—including a gargantuan maze of pipes, valves, and other controls—is enclosed in a containment shell made of concrete to prevent radioactivity from escaping from the plant in the event of a major leak. This is the building the public sees from the outside when it looks at a nuclear fission power plant, which has been designed with the intent to withstand the onslaught of floods, earthquakes, and most plane crashes.

Some people are comforted by the massive size of a nuclear-power plant. They believe this bulk carries the connotation of safety. Others believe it illustrates the dangers within. In any case, tons of steel and concrete are required to contain radioactive poisons whose lethal dangers are measured in millionths, even trillionths, of a gram.

Nuclear plants routinely release radioactivity into the water and air during their normal operations. The federal government contends this discharge is an acceptable risk. Some nuclear critics argue that the cumulative effect of this radioactive discharge may be sowing thousands of cases of cancer, leukemia, and heart disease.

Thermal pollution from a nuclear plant can be dangerous too. It can add excessive heat to the atmosphere, thus contributing to the attack on the delicate balance of the earth's temperature and climate. Thermal pollution can also harm rivers and lakes. The water that absorbs the waste heat produced by a reactor must be cooled before the water is returned to its source. A 20-degree rise in the temperature of a river or lake can damage the marine ecology and trigger massive fish kills. At Peach Bottom the state is requiring the utility to avoid raising the temperature of the river by more than 5 degrees in order to protect life in the Susquehanna.

I've toured the nuclear-power complex at Peach Bottom a couple of times. The people who work there have a tremendous pride in the plant. And I too was impressed by the care, skill, and ingenuity that have gone into Peach Bottom. Richard Fleischman, the assistant plant superintendent, was one of the men who showed me around Peach Bottom and answered my questions. Fleischman has an academic background in chemistry and has been working in the field of atomic power for more than a decade. I found Fleischman full of team spirit, extremely conscientious, hardworking, and in love with his job. He seems completely confident that nuclear power is safe. This is how Fleischman put it to me in 1974: "Look, I'm selfish like everyone else when it comes to safety. I have a wife and kids, too, you know. And we only live 3 miles from the plant. Do you think I would be working here, or that we would be living here, if there was anything to worry about?"

Thomas G. Ayers, a man with a round face, glasses, and a satisfied smile, also expresses great confidence in the safety of nuclear-power reactor plants. He is president and chairman of Commonwealth Edison Company in Chicago. Illinois, which had seven plants running in 1975, four under construction, and four more planned, is the leading nuclear state in the nation.

"I do not buy the idea that reactors are dangerous just because they have a lot of radioactivity inside," Ayers told the press in 1975. "These plants are conservatively designed, with great margins for error at every step. We have built in a whole series of multiple barriers. Also, there are various lines of defense in the safety area—systems of safety in layers. This gives me high confidence. . . . The presumption that man can't . . . make a system that won't fail . . . is sheer nonsense. . . . Whatever the problems, we could learn to master them. . . . We think the record would indicate this is a very good technology."

The other side of the coin is expressed by Dr. Hannes Alfvén from Sweden, a Nobel prizewinner in physics in 1970 for his contributions to the field of plasma physics, who is now teaching at the University of California at San Diego.

"Nobody who has visited a reactor station can avoid being deeply impressed by the ingenuity and skill which are manifest in the safety precautions," Alfvén has written. "The reactor contractors claim that they have devoted more effort to safety problems than any other technologists have. This is true. From the beginning they have paid much attention to safety and they have been remarkably clever in devising safety precautions. This is perhaps true, but it is not relevant. If a problem is too difficult to solve, one cannot claim that it is solved by pointing to all the efforts made to solve it. . . . The real question," according to this slight, shy aging scientist with a much-wrinkled face and a heavy Swedish accent, "is whether their blueprints will work in the real world and not only in a 'technological paradise.' "

The all-important task in this specialized world of technology is to keep water—or gas—circulating through the reactor fuel core to cool it and cover it to prevent the core from overheating and melting. The great danger here is that the reactor vessel might rupture (not considered very likely to happen) or that the main cooling pipe could spring a major leak. This is what most of the people in the nuclear field are worried about. A break in the main cooling pipe would permit thousands of gallons of water to escape: the tremendous pressure inside the reactor would blow all the water out through the rupture in the big pipe within 10–12 seconds.

When they're fissioning, the centers of the uranium fuel pellets heat up to around 4000° F. The cooling water keeps the surface of the pellets down around 550°. If the water blew out of the reactor and couldn't be replaced by emergency systems designed to flood the reactor vessel, the pellets would heat up excessively in less than 30 seconds. The core would begin to melt within a minute. In 10–15 minutes the core would be a molten mass weighing several hundred thousand pounds.

This huge radioactive blob would begin to eat its way down through layers of concrete and steel until it reached the earth. No one knows how far down this mass could go. Scientists and engineers call this the China syndrome. In any case, it would take years and years for this molten mass, several hundred feet or more under the ground, to cool. And once the radioactive blob reached groundwater it would probably react explosively with the water to form clouds of radioactive steam that would either shoot out through cracks in the reactor building or create explosive geysers in the ground.

Another possibility was named the Apollo syndrome by nuclear critic David Comey, who is considered especially knowledgeable about reactors and nuclear-plant accidents. This could happen, Comey said, if the core coagulated to "drop into water collected below the reactor vessel and set off a phenomenon known as a steam explosion. A tremendous release of energy, it would drive the molten core straight upward like a cannon projectile—right through the top of the pressure vessel and then through the containment dome—with an exit velocity of about 350 miles per hour. The entire contents of the reactor core would then come down . . . giving off radioactive gases. As far as I know, that would be the worst conceivable accident. . . . This is where things get gruesome because people would get radiation sickness and begin dying."

The barrier standing between us and these scenarios is the emergency core cooling system—known as the ECCS—which has been designed to send standby coolant into the reactor vessel if the primary system fails. It would have to function effectively within seconds—a minute at the outside—after that the situation would be beyond human control.

One of these safety systems would shoot water into the reactor under pressure. Another would pump it in. Unfortunately, the core could become hot enough in a matter of seconds to turn the incoming emergency cooling water into high-pressure steam; and this in turn could exert a counterpressure blocking the rest of the water from coming in. The pumps might not begin to work fast enough in the first place. Also, the federal Advisory Committee on Reactor Safeguards said emergency cooling water entering the hot reactor vessel might create a thermal shock that could crack the heavy shell of the vessel.

Moreover, no one knows for sure what would happen to the long thin fuel rods in a sudden loss-of-coolant accident. The rods could be subjected to a tremendous surge of heat from their fuel pellets during the split seconds between the time the primary coolant blew out and the emergency cooling water came rushing in—if it would, if it could. And, if the heat bent or swelled the fuel rods during this momentous, seconds-long interval, the closely stacked rods could block the flow of coolant that they would so desperately need to keep from melting.

There are many other uncertainties pertaining to the effectiveness of the ECCS. Dr. Henry Kendall of the Union of Concerned Scientists has noted, for example, that there are "10,000 or so heat exchanger tubes" in a pressurized-water reactor's steam generators, then said: "Analysis has shown that if only a handful of these tubes ruptured—perhaps eight or ten—in the brusque circumstances of an accident, the core reflooding would be entirely stalled and the reactor core would

surely melt. The possibility that this small number of
tubes, representing less than 0.1 percent of the total
number, might rupture is now believed to be very real.
In part, this is associated with extensive corrosion-
thinning of steam generating tubes with subsequent
leakage, which have been plaguing a number of reactors."

Kendall backed up his point with the reminder that
one of these tubes ruptured spontaneously in 1975 at
the Point Beach nuclear plant near Two Creeks, Wisconsin.

In 1972 the American Nuclear Society said, "It
is believed that the chances of a loss-of-coolant accident
are acceptably low for now." The society also said,
"It is quite unlikely" that the ECCS would fail. However, in 1970, the Aerojet Nuclear Company ran a
series of emergency core cooling system tests in Idaho
for the AEC using a 9-inch pilot model reactor core.
All the tests failed. In 1971, an Aerojet Nuclear report
on the tests said it was "beyond the scope of currently
used techniques and . . . some areas of present engineering knowledge" to predict what might happen in a
loss-of-coolant accident. The failures of the minitests,
incidentally, were not released to the public until nuclear critics learned of them and forced their disclo-
sure. This made the critics charge that the AEC was
once again trying to cover up some of the hazards
of nuclear power. And the notorious fire at the Browns
Ferry nuclear plant—the world's largest—indicated
that a simple flame from one burning candle could
cripple an emergency core cooling system.

The Tennessee Valley Authority's Browns Ferry
plant is about 10 miles northwest of Decatur, Alabama.
The two giant reactors were generating at full power
on March 22, 1975. It was a balmy Saturday. Workers were resealing cables at the nuclear plant. A candle
was lit to test the air flow in the cable spreading room
under the reactor control rooms. The flame was sucked
into the cable room and the wiring caught fire. The
blaze, which began at 12:51 p.m., burned for seven

hours. In a single stroke it impaired most of the ECCS, the reactor isolation system, and the remote control for several vital pumps, valves, and generators. Fortunately, there was no break in the main cooling pipes. And it was also fortunate that personnel at the plant were able to shut down the reactors manually. The Nuclear Regulatory Commission said it took 15 hours to establish "normal shutdown cooling" for one of the two reactors at Browns Ferry. Some critics noted that the Rasmussen report claimed that an accident of this kind would happen once in a billion reactor years.

Nucleonics Week, a publication catering to the industry, called the Browns Ferry fire the "most serious commercial nuclear-power incident to date." The loss was estimated at around $150 million by the TVA—over $5 million for repairs and $10 million a month to purchase substitute electricity for its customers while the reactors were down. The Browns Ferry fire also indicated that utilities around the country might be forced to spend millions of dollars—perhaps even hundreds of millions—to improve the safety of their reactor plants, those being built, ordered, or announced in 1975. And Congress and the Nuclear Regulatory Commission (which replaced the Atomic Energy Commission at the beginning of 1975) were forced to ask embarrassing questions about the standards and the inspections that are supposed to make our nuclear-power plants foolproof to serious accidents.

On February 28, 1976, the findings of the NRC's official investigation of the fire were made public, concluding that the AEC had ignored fire hazards when Browns Ferry was built in the late 1960s and early 1970s; and that the reactors had been poorly designed, built, and tested. But the plant will probably continue to operate in the future as it has in the past because the NRC study, headed by Dr. Stephen H. Hanauer, one of its senior safety experts, said it would cost too much money to make the kind of structural changes that would improve the plant's safety systems—$100–$300 million for the alterations while an additional $0.5–

$1.3 billion to purchase substitute electricity would have to be dumped on TVA's customers while the reactors were shut down.

And the critics said the NRC report ignored the unsettling implication that other nuclear plants throughout the country were operating with the same kind of flaws that were uncovered in Alabama.

Nevertheless, in spite of accidents like the one at Browns Ferry, the nuclear advocates argue that the electricity produced by nuclear plants is worth the risks. Dr. Ralph E. Lapp, physicist and consultant to the nuclear industry and writer, a sophisticated man with the comfortable looks of a favorite country uncle, once a critic of nuclear power, but now supportive, is one of the most convincing spokesmen for this point of view.

Writing in the *New York Sunday Times Magazine* in 1974, for example, Lapp said, "We must consider the question of nuclear power in this risk-benefit context." He then concluded that *if* the highest standards were enforced in the design, construction, and operation of atomic plants, nuclear fission power was "not only an acceptable risk," but the "only practicable energy source in sight adequate to sustain our way of life and to promote our economy." This view, of course, is based on the assumption that we are exhausting our supplies of natural gas and oil, that mining and burning coal would be too harmful to the environment, and that it will be many, many years before alternative sources of power such as solar energy and wind power will be available. Moreover, the nuclear establishment argues that people have always been ready to accept certain risks in exchange for the advantages and comforts that technology has to offer. And figures are cited to show that one in 4000 Americans run the risk of being killed in automobile accidents, one in 100,000 from flying in planes—but people continue to rely on the car and the airplane.

Physicist Hannes Alfvén sadly disagrees. "We are facing risks the nature of which we have never before experienced," according to Alfvén. "The consequences of nuclear catastrophes are so terrible that risks which usually are considered to be normal are unacceptable in this field." Alfvén is one of the critics who believes we can rapidly develop alternative sources of power, especially if some of the great sums of federal money that go to nuclear fission power are diverted into other areas of energy research and development.

There have also been indications that a number of people working within the confines of the nuclear establishment are worried that the benefits of nuclear power are not worth its risks. On February 2, 1976, for example, three management-level engineers in California resigned from General Electric Company's nuclear energy division, saying they were convinced that nuclear fission power was too dangerous to tolerate. The three, who had a combined total of 54 years service with the company, had been involved in the development of more than 65 nuclear plants in America and almost 30 overseas.

In announcing their resignations, which a General Electric spokesman called a "complete surprise," the three nuclear engineers, all married, with families, said they planned to live off their savings and work in the movement to defeat nuclear power in California. A spokesman for Project Survival, one of the key groups involved in California's antinuclear movement, said the three engineers "have become convinced that nuclear-power plants are not safe and they see no possibility the plants can ever be made safe enough to justify the consequences of the risks involved."

Dale G. Bridenbaugh, one of the three, had 22 years with General Electric, the company that built the reactors for the big Peach Bottom nuclear plant. The other two, Gregory C. Minor and Richard B. Hubbard, each had 16 years service with the company. A General Electric spokesman said the letters of resignation submitted by the nuclear engineers "present no fresh views or arguments but repeat emotional claims." Bridenbaugh, who described himself as "about as straight and conservative as they come," was field engineer in 1958 for Commonwealth Edison's first Dresden reactor unit near Morris, Illinois, and has been responsible

for systems to measure the performance of GE's reactors. "From what I've seen, the magnitude of the
risks, and the uncertainty of the human factor and the
genetic unknowns have led me to believe there should
be no nuclear power," Bridenbaugh was reported as
saying.

Minor and Hubbard, both of whom worked on reactor control rooms, said they were shaken by the
Browns Ferry fire. Minor, who had a key role in the
design of controls for the Browns Ferry nuclear plant,
said, "We cannot design to cover the human error, and
I am convinced the safety of nuclear reactors hangs on
the human error." Hubbard said the candle accident in
Alabama convinced him that "human error is a very
credible event," adding that he now has "a strong feeling that we don't really know what is going on inside a
reactor." Hubbard also said that "radiation is clearly
not good for life. And there's no way, I feel, with the
human error, that we can keep it away from human
beings."

The worst nuclear accident in this country occurred
on January 3, 1961, in Idaho, where three men died
when a fuel core melted and exploded at the National
Reactor Testing Station near Idaho Falls. The tragedy
was triggered by a central control rod that was withdrawn from the core of an experimental reactor, which
immediately speeded up the nuclear reaction. Since
then, workers have been injured, and two killed, at
nuclear-power plants in the United States—usually by
steam explosions, but not by radiation, according to
the nuclear establishment, which boasts that no member of the public has ever been killed or injured by an
accident at a commercial atomic-power plant.

But there have been many disturbing mishaps at
nuclear plants. The AEC, for example, said it found
safety violations in one of every three nuclear facilities
the agency inspected. These are euphemistically called
abnormal occurrences or incidents. In 1973 they were
occurring at the rate of around two a day, according to
David Comey, who keeps close tabs on these "inci-

dents," adding that the rate had jumped to four a day by 1975.

Carl J. Hocevar, one of the AEC's key experts on reactor safety, said the agency had been approving the design of nuclear plants without being assured of their safety. "The attitude was not 'Let's see if this thing works,'" the soft-spoken Hocevar said, "but 'Let's find a way to OK it.'" The small, bearded scientist left the AEC in 1974, saying he wanted to alert the public to the dangers of nuclear power, and went to work for the Union of Concerned Scientists. In his letter of resignation to the AEC, Hocevar said, "In spite of the soothing reassurances that the AEC gives to the uninformed, misled public, unresolved questions about nuclear-power safety are so grave that the United States should consider a complete halt to nuclear-power-plant construction while we see if these serious questions can somehow be resolved." Hocevar later said the AEC was using "wholly unacceptable" methods to judge the dangers of reactors.

Some of these "abnormal occurrences" stemmed from defective workmanship at nuclear plants. A welder, for example, who worked on Commonwealth Edison's Zion plant, on Lake Michigan about 40 miles north of Chicago, told an AEC safety hearing that uncertified welders had been on the job. An AEC inspection at Zion later came up with over 250 defective welds. And one plant which Commonwealth Edison operates along the Mississippi River was generating for a time without a viable ECCS because an electrician had installed a jumper cable that disabled the system.

On January 30, 1975, the new Nuclear Regulatory Commission ordered the shutdown of 23 boiling-water reactors around the country because dangerous hairline cracks were found in the reactor coolant pipes at the Commonwealth Edison plant near Morris, about 50 miles southwest of Chicago. All 23 plants were inspected to make sure there were no more cracks before they were allowed to start up again. It was the second time in nine months that nuclear plants in the

United States were closed because piping cracks had been discovered at Commonwealth Edison's Dresden plant near Morris.

Human error also plays a role in these "abnormal occurrences." A booklet put out by the federal Energy Research and Development Agency says "control-room personnel who operate the plant are highly qualified, having received extensive training and having passed comprehensive examinations to obtain the required licenses." But there have been a spate of human errors —again, for example, at the Zion plant, where a high percentage of its operators failed a requalification examination given by the AEC. On one occasion a reactor operator overpressurized the reactor vessel. Another operator caused a radioactive spill at Zion when he pumped high-level waste from one tank into another that was already full.

On June 6, 1975, 15,000 gallons of radioactive water leaked from a cooling system into the reactor containment building at Zion. The NRC blamed the leak on an open valve. "Either they were following improper procedure or procedures were ignored," according to a federal spokesman. Workers at another Commonwealth Edison plant found the auxiliary power system and backup coolant pumps under 15 feet of water because of a leaking pipe joint.

These "abnormal occurrences," of course, are not confined to Commonwealth Edison nuclear plants; they happen all around the country. The Vermont Yankee nuclear plant near Vernon, which began generating in 1972, shut down 17 times during a 19-month period. One shutdown stemmed from a fear that the control rods in the reactor might have been installed upside down.

Consolidated Edison Company, the largest utility in the country, has been plagued by costly accidents and breakdowns at its nuclear-power plants on the Hudson River, 26 miles north of Manhattan.

The Oyster Creek nuclear plant along the Toms River in New Jersey had to shut down when an operator's mistake dumped 50,000 gallons of radioactive water into the basement of the reactor building. On March 27, 1975, nearly 1200 construction workers had to be evacuated from the Northwest Utilities nuclear plant at Waterford, Connecticut, because of a radioactive water spill. Some of this contaminated water mixed with clean water and was emptied into Long Island Sound.

Radioactive gas has also escaped into the environment from nuclear plants. The Metropolitan Edison Company plant on the Susquehanna River, 29 miles north of Peach Bottom near Harrisburg (the capital of Pennsylvania), has had at least two noticeable gaseous leaks. Spokesmen for the state Department of Health and the utility insist the radioactive gas leaks have not endangered the public health and safety. But environmental groups in the Harrisburg area warn they could be serious.

And there is always the danger of earthquake faults. According to the American Nuclear Society, "The sites of nuclear-power plants are carefully chosen after much study to avoid areas of high earthquake likelihood and active fault zones. . . . Nuclear-power plants are not sited in fault zones." But the Virginia Electric Power Company built a massive four-reactor plant on the North Anna River on top of a geologic fault. The utility apparently knew about the fault while the plant was being built. Independent geologists first reported the fault to the utility in 1970. In 1973 the utility submitted a final safety report to the federal government, stating, "Faulting of rock at this site is neither known nor is it suspected." In 1974, the Atomic Safety and Licensing Board gave its approval to the North Anna site.

The North Anna Environmental Coalition, an active citizens' group based in Charlottesville, which had been fighting the North Anna plants largely on the earthquake issue, really began to raise hell when the site was OK'd. Finally, in April 1975, the federal licensing board ruled that the utility had made a dozen "false statements" between 1970 and 1973 concerning the faults at the North Anna site. The government fined Virginia Electric $60,000—the maximum amount under the circumstances. And an employee of the U.S. Geological Survey, in an internal memo, said: "I would keep my fingers crossed, and would not want to live near North Anna." Faults or not, many nuclear critics cynically believe that it would be impossible to stop the North Anna reactors from operating because so much money and time have been invested in them. Certainly money plays an important role in the whole controversy over nuclear power. By 1975, according to an important spokesman for the nuclear industry, more than $100 billion had been invested in nuclear fission power in the United States. This is a lot of money— representing a tremendous amount of pressure in behalf of nuclear power, according to the critics of atomic energy.

There are also tremendous radioactive pressures inside every operating nuclear fission power reactor plant. Pressures from within; pressures from without. Nuclear-power plants like the one at Peach Bottom, Pennsylvania, may sit in rural tranquillity, but there is a technical and moral storm brewing around every one of them.

IV
The Controversial Roots

There's a great mound of radioactive sand in the West. It could be a tragic monument to the history of the nuclear-power controversy. It's the refuse from mining uranium. And it sits in Grand Junction, Colorado. Giant sprinklers keep working here day and night to water the 55-acre mound. Health officials are trying to keep it damp in an effort to keep the low-level radioactive material from blowing away on the wind, or from sliding down into the Colorado River, which supplies people in the Southwest with water to drink and irrigate their fields.

Many, perhaps most of the people in Grand Junction don't seem very concerned about radioactivity from the great mound, or by the fact that this radioactive sand has been used as fill in the foundations of thousands of buildings in the community. But others, most of them outsiders, warn that the radioactive sand—called tailings—represents a real health threat to the people of Grand Junction and other residents of the West and Southwest.

Grand Junction seems like a nice place. The city, which has some 27,000 residents, sits in a valley in western Colorado, and is the county seat of Grand Mesa County. It's the kind of community where people smile at strangers and say hi to most everyone. On June 20, 1975, one of the days I was there, the Kiwanis Club of Grand Junction was holding its annual Pancake Day breakfast on a roped-off section of Main Street in the center of town. The morning was crisp and sunny; the air was clear and dry. Men in cowboy boots and hats, their wives and children crowded the tables that were spread out across the street. The company was good and so was the food.

Grand Junction takes its name from the nearby junction of the Colorado and Gunnison rivers. The valley is bracketed by a great picturesque rust-colored plateau, the Grand Mesa, on the east and the rugged gray-brown Littlebook Cliffs on the north. The Colorado River runs wide and muddy. There is a high wire fence between the town and the 55 acres of off-limits land along the north bluff of the river that hugs Grand Junction on the south. Rocks have been piled on the sloping bank to keep the sandy radioactive material on the bluff from sliding down into the river and washing away. The company that owns this property will have to guard it and "maintain it forever," according to Gustav A. (Bud) Franz, a close-lipped official for the state health department stationed in Grand Junction.

Grand Junction sits in the middle of the uranium-rich Colorado Plateau. The nuclear arms race, which began in the late 1940s, touched off a uranium rush in the area. Prospectors armed with Geiger counters swarmed across the Colorado Plateau and the Wyoming Basin. By 1960 there were 1000 active uranium mines spread out over nine states in the West and Southwest to feed the arms race and nuclear reactor power plants. The ore from the mines was hauled away to the mills where the uranium was extracted from the raw rock as a powder called yellowcake. And the tailings that remained behind—this fine-grained, gray, radioactive refuse—piled up outside the mills.

A lot of people in Grand Junction were caught up in the uranium rush. The Amax Uranium Corporation built a mill—now gone—in town along the river where the sprinklers now play. And people went to work in the mill and the uranium mines within a 100-mile radius of Grand Junction.

The Amax mill operated from 1952 to 1966. The tailings kept piling up: the company had no use for them. And most of the people who knew the tailings were radioactive assumed that they were harmless. The tailings were simply there for the taking.

Grand Junction had a building boom in the 1950s and 1960s. Contractors came to the mound for free fill, and, along with the general public, hauled over 300,000 tons of this radioactive sand away. They packed it around sewer pipes and underneath sidewalks and streets. They used it in the foundations of homes and schools and banks. People put the sandy stuff in their gardens to combat the clay and alkali in the soil of Grand Mesa County, and they used it to level their lawns. About 3000 homes, 15 schools, and countless lawns were built on top of sandy radioactive material in Grand Mesa County.

Then, in July 1966, a U.S. Public Health Service official and one from the Colorado State Department of Public Health together discovered that some of the buildings in Grand Junction had "elevated" radioactivity levels. The source was traced to the mound of tailings along the Colorado River. By August 1966, the state health department put a ban on the use of this radioactive sand from the mound outside the Amax mill; and the people in Grand Junction heard the site would have to be quarantined forever.

Health officials insist these uranium tailings are only "slightly radioactive," and hence do not constitute an undue danger for the public health and safety. But some scientists warn that people who live or work or play or study in the presence of uranium tailings—especially people in enclosed quarters—are being exposed to some of the conditions that exist in uranium mines.

First there's radioactive uranium oxide dust to contend with. Then, as uranium decays, it forms radioactive radium, which when it decays gives off a radioactive gas called radon. When radon decays, it in turn forms and releases descendant products called radioactive daughters. These include microscopic particles of radioactive polonium, bismuth, and lead, which can be inhaled into the lungs with particles of dust to irradiate the cells.

Many scientists believe this low-level radiation is responsible for the high rate of lung cancer in uranium miners, which, the U.S. Health Department began to warn in 1957, could develop into an occupational disease. Therefore, people in the vicinity of uranium tailings, particularly people in enclosed buildings without proper ventilation, or where rugs aren't vacuumed and floors aren't washed often enough, could be exposed to relatively high concentrations of leaking radioactive radon gas and its radioactive products—and also in some cases to uranium oxide dust.

Most of the people in Grand Junction, especially those with a substantial economic interest in the county, want to believe they have a condition that can be lived with without undue concern. But there is new evidence to indicate that low-level radiation from uranium may be more dangerous than anyone in the early 1970s imagined it could be—not only in Grand Junction, but throughout the country.

Some of the radioactive particles produced by the slow decay of uranium are turned into insoluble particles which can be inhaled into the lungs to lodge in tissues and irradiate the cells. Some advanced researchers now believe that these insoluble low-level radioactive particles may be one of the greatest threats around to our health and safety—especially for cigarette smokers.

If the evidence becomes as strong as its supporters are confident it will, respirable-sized insoluble particles that emit alpha radiation—long dismissed by many scientists as an innocuous form of radiation insofar as the health and safety of the general public is concerned—could be one of the leading contributors to the epidemic incidence of cancer and heart disease in the United States. This could be especially serious in urban areas with a lot of air pollution, or communities such as Grand Junction where there is a lot of uranium in the vicinity.

Time will tell. But time has been on the side thus far of the scientists who have been prone to be most concerned about the dangers of radiation and nuclear power, rather than those who have tended to minimize them. All the implications about the risks of cancer and other disease from natural insoluble particles that emit alpha radiation are magnified when it comes to plutonium 239—the man-made, alpha radiation-emitting by-product of the nuclear fission process, which we'll explore in the next chapter.

In addition to the risks which inhaled radioactive particles represent, some scientists are also concerned about radioactive radium 236 and lead 210 from uranium tailings in drinking and irrigation water, through which this radioactivity could find its way into our bones and concentrate, to increase the risk of human cancer, leukemia, and cardiovascular disease.

In Grand Junction, the state and federal governments, contending on the one hand that the uranium tailings are not dangerous, but apparently concerned about leaking radon gas on the other, began tearing buildings apart throughout the county, then reconstructing them, to remove the tailings and cart them away.

The state health department in Colorado and the federal Energy Research and Development Administration have called this a "health precaution." By 1976, almost $11 million—25 percent from the state, the rest from the federal government—had been spent to remove radioactive sand. Sometimes it cost $25,000 to remove a few bushels of tailings from a $35,000 home and then reconstruct the building. And removing the tailings from the schools in Grand Mesa County has been given top priority.

"A few supersensitive people from here moved away to Denver and other places," I was told by Patrick A. Gormley, the tall, lanky president of Mesa Federal Savings in Grand Junction. "But most people were concerned their houses wouldn't sell," he said, referring to the state-federal program to remove the uranium tailings. "Ninety-five percent of the people feel it's a waste of taxpayers' money, but they're for it if it's going to protect property values.

"Most of the doctors in town told the public there

was no danger from these uranium tailings," Gormley explained, "so the health hazard became secondary to the economic one." Gormley, however, suspects that low-level radiation from radon products may be more dangerous than most people believe. "My sister died of skin cancer," he said. "My father died of leukemia. And people tend to worry about diseases close to their family."

Geiger counters found uranium tailings in Gormley's home where he lives with his wife and three sons, and in the foundation of the Mesa Federal Savings building. The government paid $14,000 to remove the tailings from the Gormleys' $45,000 home. But the government's remedial action program does not cover commercial buildings, so the Savings and Loan is paying for the tailings removal work out of its own pocket.

Bud Franz, the administrator of the state tailings detection and removal program in Grand Junction, would not hand out any personal opinions about the problem or the work; but he did offer statistics, and was reluctantly willing to tell a writer what other people thought. "The general reaction here is to kinda wait and see what happens," Franz said. "And comments run the gamut of public opinion. But the majority of people want to take advantage of the remedial program and have the tailings removed from the homes and the public buildings such as schools in the area."

I went to the offices of the *Daily Sentinel* in Grand Junction, where I talked to Alice Wright, a veteran reporter who covered the tailings story, and editor Barclay Jameson. "People here couldn't get excited—people who worked in the uranium mines and the mills," Ms. Wright said. "The radiation levels in the tailings were so low they were hard to measure."

"Let's face it," Jameson said. "Uranium has been good to a lot of people around here."

The uranium mill came to Grand Junction when the AEC was buying material at high price levels in order to promote the mining and processing of uranium. When the AEC stopped buying uranium in the

1960s and let the price drop, the boom busted. Now there is a chance that new uranium money may flow into Grand Junction again. The federal government, faced by dwindling supplies of natural uranium for nuclear plants, may open the nearby Uncompahagre National Forest—full of natural treasures, like the Colorado National Monument—to commercial uranium mining.

The *Daily Sentinel* gave me Jim Temple's name, among others, and I went to see him. He is the general manager of the Public Service Company—the telephone utility. Jim Temple is a big, ruddy Westerner with an easygoing manner, who lives with his wife and seven children (four boys, three girls, aged 12–21) in a big adobe stone house that looks like a hunting lodge on one of the bluffs overlooking Grand Junction.

"I'm not worried about radiation," he said. "My wife doesn't worry either. We worry more about our kids driving around town in a car than we do about these tailings. But it really puts a stigma on the value of a house," he admitted. He said it was going to cost about $37,000 in government funds to remove the uranium tailings from the $70,000 Temple home. "The whole thing has been blown way out of proportion," Temple said.

Jim Temple kept repeating that he was "just not all that concerned. I've talked to the doctors," he said, "and they don't give me any cause for alarm about uranium." He also said his work with the phone company had brought him into contact with people who have had a part in the building and operation of nuclear-power plants. "I know they're pretty damn safe," Temple told me. "The human body is a pretty marvelous thing," he added. "It has a potential of building up a tolerance to almost anything. Perhaps the body can absorb the little bit of radiation that's around here, build up a tolerance to it. Besides, I'm a fatalist. We've all got to go someday; we're all going to die of something."

One of the doctors Temple—and most everyone else in Grand Junction—relied upon for advice about the tailings was Dr. Geno Saccomanno, a pathologist at Grand Junction's St. Mary's Hospital who is widely known for his work on lung cancer. "We have seen nothing to indicate that there have been any congenital abnormalities in Grand Junction because of the tailings," Dr. Saccomanno told me. "No increase in the incidence of cancer here, as far as we can see."

But other scientists note that it usually takes a long time for the symptoms of cancer to appear. And most victims of cancer die when they're older. Less than 1 percent, for example, of the people who die of lung cancer are under age 40; less than 2 percent die before 50. Then the death curve begins to rise. Eleven percent die between 60 and 65 years of age; 59 percent die when they're between 70 and 80 years old, which means that attempts to correlate the effects of radioactivity from uranium tailings could be tenuous.

Dr. Saccomanno has been cited for early detection of lung cancer, especially among uranium miners. Hearings before the Joint Committee on Atomic Energy in 1967 reported that 65 percent of the uranium miners in the Colorado Plateau were exposed to "concentrations" of radon gas and "radon daughters," "substantial portions" of which are "retained in the lungs," according to the hearing report, which said "radon gas will largely accumulate in the fatty tissues" to irradiate the cells.

The Joint Committee concluded that it was impossible to state with certainty how dangerous radon and its by-products are. But data collected in the 1930s established that 30–50 percent of the men who worked the uranium and radium mines before the Second World War in what was then southern Germany, now a part of Czechoslovakia, died of lung cancer, which was attributed to radon gas in the mines. And 400 out of 6000 American uranium miners in the West died of lung cancer by 1975, with more cases of miners with lung cancer being reported as time goes by.

In 1957, Dr. Saccomanno began to examine samples of the sputum (spit) from uranium miners. By 1975 he and his colleagues at St. Mary's Hospital had studied over 75,000 specimens of sputum. During the course of their research "we watched how cancer develops in the lungs," Dr. Saccomanno told me, claiming that carcinogenic cells in sputum can be discovered under a microscope long before they can be spotted in X rays. With this knowledge the cancer victim and the doctors can keep an eye on a developing tumor until it reaches the point where it is large enough to try surgery. Unfortunately, lung cancer is usually fatal. Ten percent or less of all the people who get lung cancer in the United States are saved, according to the American Cancer Society.

Dr. Saccomanno was convinced when I talked to him in 1975 that cigarette smoking, rather than radon gas and its by-products, was primarily responsible for the high incidence of lung cancer in uranium miners. "Smoking changes the metabolism of the lungs," Dr. Saccomanno told me when he was telling me about his experiences dissecting the lungs of miners over the years who had died of lung cancer. "It causes a transformation in the lining of the lung, which undergoes further changes—and then the lung becomes subject to radiation."

When I met him, Dr. Saccomanno had been with St. Mary's Hospital for 26 years. Short, strong, and balding, with a slight limp and Italian immigrant roots, Dr. Saccomanno had worked his way through medical school, and like many first-generation Americans he seemed to have a fervent belief in the work ethic and business institutions. "People who cry that radiation is harmful are not being fair to uranium mines and the companies that own them and run them," Dr. Saccomanno complained. "This country has to have atomic energy in order to survive. The risks are minimal. People who talk about radiation are not being fair to the nuclear plants. The real culprit," Dr. Saccomanno insisted, "is cigarette smoking, which is killing about 85,000 people a year from lung cancer. And cigarette smoking is behind atherosclerosis too."

Dr. Saccomanno told me his research work was supported by the ERDA and the U.S. Public Health

Service. "ERDA is very interested in finding out the facts," he said. But the nuclear critics wonder how willing ERDA would really be to accept—much less publicly acknowledge—new evidence on the risks of low-level radiation from natural elements in the environment, and from the radioactive products produced by nuclear plants and nuclear-bomb tests—especially the risk that plutonium represents.

I've talked to some of the scientists who work directly or indirectly for ERDA. They are staunch believers in nuclear power, or at least say they are; and many of them do tend to resist new ideas and data about the dangers of radiation, or anything else for that matter, that calls the desirability of nuclear fission power seriously into question.

One wonders, too, about the many tons of radioactive uranium tailings that have piled up in the basins of the Colorado River around towns such as Grand Junction. Some of the tailings along the banks of rivers and streams must have been washed away toward reservoirs such as Lake Mead in Nevada, which supplies Los Angeles with drinking water.

And all the while people in Grand Junction were becoming part of the history of the nuclear-power controversy, nuclear power in the United States was developing into a multibillion-dollar industry. There were a number of reasons for this growth. One was guilt; people who developed the atom bomb during the Second World War, and some who supported the arms race that followed, were desperately anxious to believe that the destructive uranium atom could be harnessed in peacetime as an energy panacea. Many of these believers thought that nuclear power would be safe, efficient, and cheap—so inexpensive, in fact, that the electricity produced might be too cheap to meter. A number of giant corporations, including Westinghouse and General Electric, which had made tremendous investments in talent and capital to obtain atomic military contracts, were anxious to pursue new sources of financial return in the nuclear field. So they too began to claim that nuclear fission power was the only practicable answer to our energy needs in the immediate future.

There was also a fear in the private power sector that the federal government would go ahead and build nuclear-power plants if the private utilities did not. So the electric companies, many of them worried about the dangers and uncertainties of nuclear fission power, felt compelled to build atomic plants out of fear that they would be left behind at the dawn of a new age of electrical power.

One of the people who loom up when we look over the history of nuclear power is Albert Einstein, who in 1905 literally changed the world we live in with his special theory of relativity. Einstein proved that matter—made up of submicroscopic atoms, a word that comes from the Greek, meaning "that which is invisible"—could be converted into energy. Therefore, certain atoms could be split—or fused—to convert part of their mass into energy. Einstein's theory set off ripples of research, most of them in Europe, into the mysteries of the atom. Nuclear concepts were already fairly well developed by the time fascism drove many prominent scientists and scholars to the United States—among them Einstein, a German Jew; Enrico Fermi, an Italian; and Leo Szilard, a Hungarian.

They continued their research work in the United States while they anxiously watched the war spread over Europe. And they were afraid that the Germans under Hitler might be the first to develop an atomic bomb. In 1939 Szilard got Einstein to write a letter to Franklin D. Roosevelt, informing the president that it might be possible to build a nuclear bomb. So the United States began to commit money and personnel to try to develop this awesome weapon.

In August 1942 the Manhattan Engineer District Project was formed. This was the code name for the wartime predecessor of the Atomic Energy Commission; its mission was to build the bomb and keep it secret. In December 1942 a team under the direction of physicist Enrico Fermi, using a pile of uranium moderated by graphite, produced the world's first nuclear chain reaction in an abandoned squash court under Stagg Stadium at the University of Chicago. By January 1943 the federal government was building atomic-bomb plants at Oak Ridge, Tennessee, and Hanford, a remote 570-square-acre tract along the Columbia River in the southeast part of Washington, where millions of gallons of high-level radioactive waste have been stored—of which thousands of gallons have leaked. The atom-bomb workers at Oak Ridge were trying to produce highly enriched uranium; those at Hanford built plants to produce plutonium from uranium.

On July 16, 1945, the world's first atom bomb was exploded in a fiery cloud over the sands of New Mexico near Alamogordo. The heart of it was a plutonium sphere no larger than a baseball. The bomb, called the Fat Man, went off with the force of 20,000 tons of TNT. People who witnessed the blast were stunned: they were afraid the spreading cloud might never stop.

On August 6, 1945, a bomb using enriched uranium was dropped on Hiroshima, Japan. Over 65,000 people perished in the blast from the explosion. Thousands more were injured by radiation poisoning. Three days later a plutonium bomb was dropped on Nagasaki. Twenty-five thousand people were killed outright. By 1975, the Nagasaki death toll from the bomb was over 48,000. World War II was at an end.

The following year, on August 1, 1946, the Atomic Energy Act was passed, creating the Atomic Energy Commission. Its first chairman was former Tennessee Valley Authority czar David Lilienthal, who later felt compelled to express some grave misgivings about the dangers of nuclear fission power.

On December 31, 1946, the control of atomic power in this country was taken away from the military's Manhattan Project and placed in the civilian hands of the AEC. The cold war was on. So was the atomic-bomb race, in its beginning phases, which created the uranium rush around Grand Junction and other towns in the West and Southwest.

The Joint Committee on Atomic Energy was created, with unprecedented authority in Congress—because of the fear and secrecy then prevailing—to exercise control over the AEC programs, both for war and for peace.

The U.S. Navy, under the well-publicized prodding of Admiral Hyman G. Rickover, began to push the development of the nuclear submarine, which in turn planted the seeds for the growth of nuclear-power plants in this country. The federal government was paying the bills. Westinghouse and General Electric were swept up in the submarine program; they sensed that commercial atomic power could become a highly profitable venture.

In 1955, the U.S. submarine *Nautilus,* using a pressurized-water reactor, had its successful maiden voyage, which spawned the development of a huge American nuclear fleet. Some people began to feel that this meant nuclear-power plants were just around the corner. They were afraid the plants might be developed as a government monopoly. There was an even greater fear that the nuclear arms race would end in a universal holocaust; this led to President Dwight D. Eisenhower's "Atoms for Peace" declaration. Speaking before the United Nations on December 8, 1953, Eisenhower said, "If the fearful trend of atomic military buildup can be reversed, this greatest of destructive forces can be developed into a great boon, for the benefit of mankind."

The next year, in response to pressures from industry, the AEC, and the Joint Committee, Congress passed the 1954 Atomic Energy Act, which permitted the private sector—under the jurisdiction of the AEC —to own and operate atomic-power plants and their related facilities, among them plants to process nuclear waste materials. Meanwhile, the AEC was beginning to develop into one of the most controversial federal agencies in the history of the United States. It was charged with two responsibilities: promoting the growth of commercial nuclear power, and protecting the public health and safety from its dangers. Some concerned citizens were soon to complain that they were unable to discern where the AEC's promotional zeal ended and its regulatory function began.

Private utilities and manufacturing companies joined together to form nuclear-power-plant development firms. But the pursuit of nuclear power was fraught with huge financial risks and questions about the safety of nuclear fission power. Therefore, the AEC said it would subsidize, in large measure, the embryonic stages of nuclear-power plants in this country. In 1956 the AEC sweetened the bait by saying it would buy the plutonium produced by commercial power plants for the government nuclear warheads program.

The nuclear lure was getting more attractive, but it was still not tempting enough. No one would insure a nuclear-power plant: the risks of accidents seemed too great. The Joint Committee put it this way: "The problem of possible liability in connection with the operation of reactors is a major deterrent to the further industrial participation in the program. . . . Liability has bcome a major roadblock."

The AEC then commissioned its first study from the Brookhaven National Laboratory to probe the potential risks of nuclear-power plants and determine how much insurance might be needed to cover them. Around the time that the findings of the study were presented to Congress in March 1957, indicating that a major nuclear-plant accident could kill and injure thousands of people and cause billions of dollars in property damages, Herbert W. Yount, vice-president of Liberty Mutual Insurance Company, testifying at a federal hearing in behalf of the American Mutual Alliance, said in part: "It is a reasonable question of public policy as to whether a hazard of this magnitude should be permitted, if it actually exists. . . . Obviously. . . . no principle of insurance . . . can be applied to a single location where the potential loss approaches such astronomical proportions. Even if insurance could be found, there is serious question whether the amount of damage to persons and property would be worth the possible benefit accruing from atomic development."

Congress passed an amendment to the Atomic Ener-

gy Law of 1954 in September 1957 called the Price-
Anderson Act, whose stated purpose was to "protect
the public" and "to remove a deterrent to private industrial participation in the atomic energy program
posed by the threat of tremendous potential liability
claims." The 1957 act, named after the late Senator
Clinton P. Anderson, a Democrat from New Mexico,
and Melvin Price, an Illinois Democrat, both members
of the Joint Committee on Atomic Energy, attempted
to strike this balance by limiting the liability for a single
nuclear-power plant accident to $560 million, of which
$60 million, when the law was passed, would be paid
by the nuclear utilities, which would purchase their
coverage from a pool of private insurance companies;
the rest would be guaranteed by federal taxpayers'
money.

The American Nuclear Society heralded the Price-
Anderson Act as an endorsement of atomic power because the $60 million coverage "provided by the in-
surance companies is the greatest commitment they
have ever made for a single hazard." The insurance
companies have gradually increased their coverage
over the years—to $125 million by 1975, saying their
action was based on the record of safety at nuclear
plants over the years.

The act as passed was full of loopholes. It did not,
for instance, insure the manufacturers of nuclear
equipment; it simply exempted them from any liability.
The victims of a nuclear accident were governed by
the law's 10-year statute of limitations—increased to
20 years in 1975—even though it often takes 15–30
years for cancer symptoms to develop. The victims also
had to prove that a particular incident was responsible
for their disease. Unfortunately for the injured, a tumor
does not sprout a little tag identifying uranium or plutonium or other radioactive elements connected with
the nuclear process as its cause.

Dr. Chauncey R. Kepford, a chemist and nuclear
critic from Pennsylvania, years later complained that
the "act contains no definition of radiation injury."
Testifying against renewal of the Price-Anderson Act,

Kepford told the Joint Committee in September 1975 that "it seems incredible that Congress could have, in effect, created a new industry without defining what constitutes an injury from the hazard that necessitated the liability limit."

Kepford said the "crux of the problem" lay in the fact that "there is no legal definition of just what constitutes an injury due to radiation exposure. The courts are as much without guidance as the public." And Kepford charged that members of the nuclear establishment "have not even acknowledged the existence of the problem."

Another critic, former Pennsylvania insurance commissioner Herbert S. Denenberg, a short, pugnacious, consumer-oriented gadfly, later cried, "It may be that nobody but God could write the insurance policy we need on nuclear plants." Denenberg estimated in 1973 that an insurance company would have to charge an annual premium of $23.5 million in order to make a profit in light of the potential damages that could accrue in the event of a major nuclear-power-plant accident.

There was another nuclear milestone in 1957 along with the enactment of the Price-Anderson Act. This was the year the first atomic-power plant in the United States began to generate electricity on the Ohio River near Shippingport, Pennsylvania, about 25 miles northwest of Pittsburgh. Jointly owned by the AEC and the Duquesne Light Company, and operated by the utility, the plant featured a 60-megawatt pressurized-water reactor (upgraded to 90 megawatts in 1965)—a much larger model of the type used to power nuclear submarines.

Westinghouse and General Electric at that time practically had the reactor field to themselves. Because both had gone into nuclear power with the development of the atomic submarine, their reactors, and the subsequent design of nuclear plants in America, were based on the demands of military technology—which did not make very efficient use of uranium fuel. Westinghouse had the pressurized-water model; General Electric the boiling-water reactor. And people say that both were willing to accept multimillion-dollar losses in the beginning in order to seed the utility market for their new nuclear-power-plant technology.

But even so, the utilities were still reluctant to build atomic plants. Admiral Lewis L. Strauss, an investment banker from Wall Street, a cold war hawk who had succeeded David Lilienthal in 1953 as the chairman of the AEC, was highly impatient with this foot-dragging. He wanted to put commercial nuclear power —which had been developed with taxpayers' money— into private hands.

Rather bluntly, Strauss warned the private sector that the federal government would build nuclear-power-er plants and monopolize the new industry if the utilities refused to build reactor plants. This warning began to move some of the utilities toward nuclear power. There were some utility executives who didn't need any prodding from Strauss. One of them was Walker L. Cisler, chairman of the Detroit Edison Company, who gathered 14 utilities and 7 equipment manufacturers together to form the Power Reactor Development Company in 1955—a Michigan corporation with Cisler as president. Some of the biggest private names in today's atomic-power establishment were involved, among them Westinghouse, Babcock and Wilcox, Combustion Engineering, and Philadelphia Electric Company.

Cisler was attracted by the possibilities of the liquid-metal fast-breeder reactor, the most exotic model on the drawing boards, which would create heat to generate electricity and produce more plutonium for weapons and fuel than the uranium consumed by the reactor. And so plans were drafted to build the Enrico Fermi fast-breeder reactor plant. The site of the Fermi plant was about 30 miles from Detroit, and 20 miles northeast of Toledo, Ohio, at Lagoona Beach on the western shore of Lake Erie near Monroe, Michigan, which had a population of 20,000.

Cisler was highly enthusiastic about the project, according to all accounts. The 60-megawatt Fermi reactor power plant was supposed to cost about $50 million; and Cisler was saying that by 1970 Fermi would earn $43 million for its investors by selling electricity to Detroit Edison, and another $48 million selling plutonium to the AEC for nuclear weapons. But some scientists and nuclear engineers, from the very beginning of the Fermi project, were extremely leery of the liquid sodium-cooled fast-breeder reactor. Hot liquid sodium reacts violently with water or air: it burns in air and explodes when it comes in contact with water. And the sodium from the reactor core has to be piped through water to create steam to produce electricity, so some nuclear critics were afraid that a leaking sodium pipe could produce a serious accident.

There were other fears about the breeder reactor: it is smaller and more compact than water-cooled reactors, and the neutrons in the breeder travel at speeds of 30 million miles per hour, compared to 10,000 miles an hour in a light-water unit. A breeder could suffer a runaway nuclear reaction in just a few seconds and go out of control, to possibly explode with enough force to rupture the reactor containment shield and building.

A breeder reactor could also contain hundreds of pounds of plutonium 239. Only 1 pound of plutonium, properly dispersed into the atmosphere from a major nuclear accident, could cause an epidemic of lung cancer.

The Fermi fight—over its construction, safety, and the partial meltdown of its fuel core in 1966—developed into a classic confrontation that has given shape to many aspects of the nuclear-power controversy that persist to this day. On June 28, 1956, AEC Chairman Lewis Strauss announced that he would attend groundbreaking ceremonies on August 8 for the Fermi plant. The next day, June 29, the AEC's Advisory Committee on Reactor Safeguards disclosed that it had submitted a report to the AEC on June 6, 1956, saying the committee "believes there is insufficient information at this time to give assurance that the PRDC [Fermi] reactor can be operated at this site without public hazard." The critics say Strauss knew about this critique, but that he deliberately chose to ignore it when he said he was going to Michigan in August for the Fermi ground-breaking ceremony.

G. Mennen Williams, then governor of Michigan, asked for assurances about the Fermi breeder plant. On July 17, 1956, the AEC replied—as if it were a military secret—that it would be "inappropriate" to let the governor know about the alleged dangers of the proposed reactor. And the critics charge that the AEC has consistently and persistently tried to hide the dangers of nuclear-power plants from that day on.

On August 3, 1956, the AEC issued Power Reactor Development Company of Detroit a "provisional" construction permit for the Fermi reactor plant. When this happened, Senator Clinton Anderson, chairman of the Joint Committee on Atomic Energy, wired a warning to Governor Williams of Michigan, accusing the AEC of "star chamber" proceedings for issuing a building permit in the face of "grave doubts" about the safety of Fermi.

Anderson also predicted the tenor of licensing proceedings when he said, "This construction permit . . . sets a dangerous pattern for the early stages of AEC regulative and quasi-judicial activity. . . . The governor has apparently been precluded from being heard or participating in any decision to build this reactor in the state of Michigan." Anderson added that the AEC may have "violated established legal principles by the confusion of its development and promotion functions with its regulative and quasi-judicial functions."

The next day Congressman Chet Holifield, a Democrat from California and another member of the Joint Committee who later became an ardent long-standing champion of the atomic establishment, sent telegrams protesting the decision to build the Fermi plant to President Eisenhower and Governor Williams. In his wire to the president, Holifield pointed out that the reactor-safeguards people "reported unfavorably on the design of this reactor and pointed out inherent hazards," and said he hoped Eisenhower would urge the AEC to "rescind the construction permit in consideration of health and safety" once the White House had the "full facts."

The wire to Williams urged the governor to "take such steps as you deem necessary to protect life and property in the state of Michigan." And Holifield said he was expressing the fears of the Joint Committee when he added, "While the AEC permit is 'allegedly conditional,' I am convinced that once large sums of money are spent for construction, pressures to have reactor operated will be overwhelming. Therefore must urge every possible step to stop construction." But AEC Chairman Lewis Strauss was determined to build the Fermi reactor plant, and his will prevailed.

On August 29, 1956, after the so-called conditional AEC building permit had been issued, the AEC opened public hearings on the construction of the plant. The next day three labor unions, with the full backing of the AFL-CIO executive council, and led by the United Auto Workers, intervened on the grounds that the Fermi breeder was "inherently unsafe" and was therefore a danger to union members who worked in Detroit. Burly Leo Goodman, who was the late Walter Reuther's energy adviser, directed the labor fight against the Fermi reactor. The hearings lasted for 11 months. Goodman kept insisting that a breeder reactor "might convert itself into a small-scale atomic bomb."

The developers of Fermi brought in consultants to refute this allegation. One of them was Dr. Hans A. Bethe, Nobel laureate, expert on nuclear weapons, and professor of physics at Cornell University, who flatly testified that the reactor in question could not explode like a bomb; and that he was "confident that the Fermi plant can be operated safely and will give satisfactory results."

Goodman complained that the hearing before the three-member AEC licensing board was a farce from the beginning because it was up to the intervenors to prove the reactor was dangerous, rather than to the AEC and its developers to demonstrate that it was safe. "And the only way we could do that," Goodman told me in 1975, "was to wait for an accident to happen."

On December 10, 1958, the AEC licensing board ruled that the Fermi plant could be built. The unions went to court. On June 10, 1960, the U.S. Court of Appeals in Washington, D.C., in a 2–1 decision, halted construction of the nearly completed plant, ruling that the AEC had to guarantee the safety of the Fermi plant before the agency could issue a construction permit.

The developers appealed up to the Supreme Court, which, on June 12, 1961, ruled in favor of the AEC's two-step licensing procedure—which was to issue a construction permit even though all the safety questions had yet to be answered, then an operating permit once the plant was built and all the questions concerning it had apparently been resolved to the satisfaction of the AEC.

In a biting dissent, Justices Hugo L. Black and William O. Douglas called the AEC's licensing procedure "a lighthearted approach to the most awesome, the most deadly, the most dangerous process that man has ever conceived."

Leo Goodman responded: "To locate this experimental plant . . . so near population centers is as reasonable as trying to control a 10-ton truck with untested brakes in a congested city street." Some people call Leo Goodman the godfather of the fight against nuclear power in the United States. Certainly, he was one of the first intervenors. And he has confronted the multibillion-dollar nuclear establishment on hundreds of occasions and at countless government hearings. Goodman still insists that the Fermi breeder reactor could have exploded—that its fuel rods could have melted and gone off with the force of 100 tons of TNT—the equivalent of a small tactical nuclear weapon. He also says the "AEC was dishonest and tricky from the word go. Watergate was nothing new to me," Goodman said in 1975. "I've been living with the dirty tricks of the AEC right from the beginning."

Meanwhile, the costs of the Fermi plant kept going up. In 1955 they were estimated at $50 million; by mid-1961 they were $82.6 million; and by the time the Fermi reactor was ready to start up for the first time in 1966 about $120 million had gone into the plant.

There had also been some ominous warnings from the West. On November 29, 1955, the fuel core melted in an experimental fast-breeder reactor called the EBR-1 at the National Reactor Testing Station in Idaho. In 1961 the fuel core in the SL-1 reactor at the test site melted and the unit exploded, killing the three men working with it.

In January 1966, the Fermi plant began to generate power. But trouble developed in the crucial steam generator, where the hot liquid sodium from the reactor was pumped through heat exchangers filled with water to produce steam for electricity, and the reactor was shut down for months. On October 4, 1966, Fermi was ready to try again. The control rods were slowly withdrawn, releasing the flow of neutrons to start their nuclear chain reaction. The reactor ran for a while. Cautiously, the power was slightly increased. Then the reactor was shut down for the night.

The next day, at 1:45 p.m., the control rods were withdrawn again. The chain reaction began. At 3:00 p.m., when the reactor was producing at only a fraction of its capacity, there were warning signals in the reactor control room. At 3:05 p.m. the operator switched from automatic to manual control of the control rods. Suddenly the temperature readings on the control board showed there were two dangerous hot spots in the core of the reactor. Radiation alarms had begun to sound in the reactor building. Fortunately, there was no one inside the building at the time. Automatic devices sealed it off. At 3:20 p.m., on October 5, 1966, all six control rods were pushed into the fuel core as far as they would go to "scram" the reactor by absorbing the flow of neutrons.

People at the plant and in nearby Detroit took a

deep breath and prayed—for months. At one point, it was feared that Detroit with its 1.5 million inhabitants might have to be evacuated. It was obvious that part of the fuel core had melted. The crucial question: Would enough of the fuel fuse together and melt to set off a chain of events that could cause an explosion? If so, the reactor vessel and containment shell could rupture, releasing radioactive steam and debris to the wind.

Workers at the Fermi plant were terrified; they were afraid they could touch off an explosion if they poked around inside the reactor to find out what had gone wrong. It was months before they were willing to risk lowering a periscope into the reactor to see what had happened. Then they learned that two of the fuel core assemblies had melted; two others had been damaged.

This was more serious than the "maximum credible accident" Power Reactor Development Company of Detroit had said could happen—which was one fuel bundle melting. The Fermi accident involved two. The company also said that "the reactor would probably be shut down automatically" if a fuel bundle began to melt; but what actually happened is that the reactor had to be manually shut down. A preliminary investigation indicated that metal—at first thought to be some beer cans that had been left behind inside the reactor by some construction workers—had clogged and then blocked the flow of sodium coolant from the bottom of the reactor vessel.

It took a year to dismantle the reactor. The metal culprit turned out to be a plate of zirconium. This is what went wrong, according to Congressman Craig Hosmer, a Republican from California and a member of the Joint Committee on Atomic Energy: "Late in the construction of the reactor, around 1950, the designers decided to cover the bottom of the vessel with a zirconium sheet as an added safeguard to prevent fuel melt from going through the bottom of the vessel. . . . They hurriedly rounded up some zirconium sheets . . .

and they were bolted on the cone. . . . At least three bolts were used in each. . . . The total cost of the sheets was about a hundred bucks."

The force of the sodium flow loosened the bolts on one of the zirconium sheets. It floated around until it finally came to rest where it could block the flow of the incoming coolant. The designers neglected to add their last-minute handiwork to the blueprints, so everyone had forgotten the metal sheets were there.

Reporter Saul Friedman covered the Fermi story for the *Detroit Free Press*. When I talked to Friedman in 1975, he was a highly respected journalist with the paper's bureau in Washington, D.C. This is what Friedman told me: "From the very beginning, Walker Cisler of Detroit Edison kept minimizing the dangers of a potential accident at Fermi. And then he minimized the accident when it did happen. He kept saying, 'the prevailing winds' would 'blow over to Canada.' "

It was four years before the Fermi plant was able to start up again. On November 29, 1972, the utility announced that the troubled Fermi plant would be permanently closed. Around $133 million had gone into it —to sell only $300,000 worth of electricity. During its lifetime the plant had only been able to operate for about 30 days—and never for more than five in succession. Detroit Edison even spent around $23 million to build an oil-fired power plant at the Fermi site to produce electricity when the reactor wasn't working.

The nuclear establishment has gone to great pains to minimize the disturbing history of the Fermi reactor plant—the partial meltdown of its fuel core, its operating problems and financial difficulties. Trying to make the best of a bad situation, an AEC spokesman later claimed that "the world learned a lot" of beneficial information to aid in the development of nuclear power from the Fermi plant. Around 77,000 gallons of radioactive sodium were at the plant when it closed. In 1975 the sodium was still there, frozen in 55-gallon barrels and large storage tanks near Monroe, Michigan. Power Reactor Development Company of Detroit said it didn't know what to do with the radioactive sodium —largely because of the cost to have it hauled away and disposed of.

The 1960s were important to the history of nuclear controversy in America—and not only because of the Fermi fast-breeder reactor. On December 10, 1962, Consolidated Edison Company, the nation's largest private utility, applied for an AEC construction license to build a 700-megawatt reactor plant in the Ravenswood district of the New York borough of Queens, across the East River from Manhattan's 72d Street.

The proposal to build a nuclear fission reactor power plant in such a populated site as New York City prompted formidable opposition, and the plan was dropped. But the controversy it stirred up—especially about the dangers of nuclear plants to populated centers—helped generate considerable interest in the budding debate over nuclear power. The nuclear proponents, in turn, had to be content with building reactor units near small communities—which couldn't mount the opposition a major metropolis could—within transmission range of population centers.

By the mid-1960s orders for nuclear-power plants were coming in, and the multibillion-dollar nuclear establishment was firmly entrenched. Its fervent, powerful members included the bureaucratic AEC, with over 6000 employees by then, whose highly paid jobs were linked to the proliferation of nuclear power plants, the congressional Joint Committee on Atomic Energy, and firms such as Westinghouse, General Electric, the Bechtel Corporation (one of the architect-engineer-design outfits that were building the plants), the Kerr-McGee Corporation in Oklahoma, which was mining and processing uranium, General Atomic, and many others. And the AEC was making the rules governing nuclear power.

The critics complained that the AEC licensing proceedings were developing into a charade. Senior AEC scientist Dr. John Gofman charged that "licensing boards went through the motions of a hearing and inevitably returned a rubber-stamp decision that a go-ahead should be given to every nuclear-power plant." But the physicists, engineers, and other scientists for the nuclear utilities seemed to sincerely believe that the AEC was an often obnoxious adversary, one which was subjecting the utilities to an endless routine of paperwork and red tape, and tedious, time-consuming meetings and hearings devoted to nuclear safety, when the utilities were already convinced they had resolved all the safety problems at their plants a hundred times over.

The nuclear-bomb tests were still going on in the 1960s; a worldwide ban-the-bomb movement had sprung up to stop them. Dr. Linus Pauling, a Nobel laureate in chemistry, and Norman Cousins, editor of the *Saturday Review,* were among those in the United States who led the fight against nuclear testing. They were concerned about the dangers of radioactive fallout. And this controversy revolved around the harm that low-level radiation from elements produced and released by the bomb could do.

The AEC, which was promoting the tests, tried to claim that this fallout was not dangerous. Then the agency said the danger was from external rather than internal radiation from fallout. And the AEC was again proved wrong. Next the AEC measured the amount of radioactive fallout released by a bomb, averaged it out over hundreds and even thousands of square miles, and concluded it was nothing to really worry about. Those opposed to nuclear testing warned that excess radiation, no matter how little, would have significant adverse health effects. And they quoted Dr. Hermann J. Muller, a Nobel prizewinning scientist, who had said, "There is no amount of radiation so small that it cannot provide harmful effects."

It was soon evident that fallout was not evenly distributed, but was often highly concentrated locally when it came down in the form of radioactive rain. It was also shown that strontium 90, iodine 131, and a few other fallout constituents were concentrating in the human food chain. They accumulated in grass, for example, which was eaten by cows to produce milk with relative concentration levels of radioactivity that were hundreds and even thousands of times as high as that in the fallout itself.

In 1963 the United States, the Soviet Union, Great Britain, and most other nations signed an agreement to stop testing nuclear bombs in the atmosphere. By this time many people were questioning the honesty and objectivity of the AEC, and this skepticism carried over into the nuclear-plant controversy, where the debate over the dangers of low-level radiation continued unabated because nuclear plants produce the same radioactive poisons that nuclear explosions do.

In the year the test ban was signed the AEC assigned two of its top scientists to investigate the dangers of low-level radiation from nuclear-power plants and report back to the agency with their findings. They were Drs. Arthur R. Tamplin, a biophysicist, and John Gofman, a nuclear medical-physicist. The two were research associates at the Lawrence Livermore Radiation Laboratory in California. Dr. Gofman, who was also a physician, was the associate director of the Livermore Laboratory from 1963 to 1969. His special fields of interest were cancer and heart disease. The two men began to explore the risks of low-level radiation. They were assisted by Donald P. Geesaman, a brilliant young physicist who through his intensive research work, along with Tamplin, later developed into an authority on the toxicity of plutonium.

Gofman and Tamplin started out with the common knowledge that there is natural background radiation all around us—in cosmic rays from the sun, in the atmosphere, and in soil, rocks, and minerals. Some of this natural radiation can penetrate our skins to irradiate our cells; other radioactive elements can collect in dust particles, which we inhale, and in food and liquids that we take into our bodies. In addition, we are exposed to man-made radiation—from X rays, television sets, and from the nuclear process (from fallout, reactor plants, and other nuclear facilities). This radiation is measured in units called rads—a measure of the energy absorbed per gram of body tissue from a particular radioactive element—or rems—roughly the equivalent of a rad for X rays, gamma rays, and beta particles, but about a tenth as much for neutrons and alpha particles—a measure of the relative biological damage to living tissues.

The Federal Radiation Council, established by President Eisenhower in 1959, held that anything above a maximum lifetime dose of some 250 rads could be dangerous. The Council estimated that the average American received around 110 millirads (1000 millirads = 1 rad) of radiation each year from natural background sources; and concluded that the average person could probably absorb an additional 170 millirads a year—from X rays, power plants, and television viewing—without undue harm. The council guidelines did not suggest that everyone would be exposed to an extra 170 millirads, or that this would be safe; only that no one working in the nuclear industry should receive more than 5000 millirads per year (a standard that has drawn increasing fire over the years from critics who charge that atomic workers are being sacrificed to the hazards of nuclear power), and that the average dose to the general public should not exceed 170 millirads above and beyond the background level.

Gofman and Tamplin used this 170-millirad figure as the base for their lengthy research; in October 1969 they presented the AEC with their findings, concluding —conservatively so, they thought—that if everyone received this additional 170 millirads of radiation per year, "there would, in time, be an excess of 32,000 cases of fatal cancer plus leukemia per year, and this would occur year after year." The two scientists claim they expected the AEC to "welcome our report on cancer-plus-leukemia risk—especially since the findings were being made available before a massive burgeoning of the nuclear electricity industry." Instead, their report touched off a furious controversy that is still going on within government circles and the nuclear industry.

The American Nuclear Society quickly derided their claims, contending their conclusions were "false," alleged they were based on "improper use of existing data" and were aggravated by "impossibility." But David L. Levin, a spokesman for the National Cancer Institute, later said, "Using different methodology . . . we computed risks to be of the same general level as those shown by Dr. Gofman."

One thing is sure: the amount of radiation—or threshold—which the regulatory bodies consider to be an acceptable risk keeps being revised downward as scientists learn more about the harmful effects of radiation. The suggestions that Gofman and Tamplin made in 1969 about the amount of radiation a nuclear-power plant should be allowed to release were endorsed by the National Academy of Sciences and finally adopted by the AEC: these were 1 percent of the old AEC allowances.

In retrospect, Dr. Gofman is somewhat bemused, and highly skeptical as a result of his brush with the nuclear establishment. "When the project started in 1963," he recalled in 1975 when we were talking in his home in San Francisco, "I didn't feel that atomic energy was as bad as some people said it was. My chief interest was cancer and chromosomes. But I was suspicious of the AEC. And I asked Seaborg [Dr. Glenn T. Seaborg, a physicist, then chairman of the AEC] if the AEC wanted the facts or a 'gloss' to excuse the operation of nuclear-power plants. Seaborg assured me they wanted the facts."

Dr. Gofman is a spry, shy, gray-bearded man who says, "I basically love working in a laboratory." In 1947, after getting his M.D. and a Ph.D. in nuclear-physical chemistry, Gofman joined the medical physics faculty of the University of California at Berkeley. In the late 1940s and early 1950s he did research work for the AEC. From 1953 to 1957 Gofman worked a few days a week as the medical director of the Lawrence Radiation Laboratory and became an associate director of the facility at Livermore, with its 5000 workers, in 1963, serving in this capacity until 1969, when he and Tamplin published their controversial findings on the risks of low-level radiation.

Tamplin had 12 assistants at Livermore when the report came out, according to Gofman, who said Tamplin began to lose them one by one until he only had one left. "And I gave the AEC good cause to fire me," Gofman said, "but they didn't want to. They were afraid they would get a lot of bad publicity if they let me go."

By 1969 Dr. Gofman was widely known in scientific circles and had an exemplary reputation. He was codiscoverer of two protactinium isotopes and two uranium isotopes, including uranium 233, which can be used to generate nuclear power; had coinvented processes to separate plutonium; and had gathered awards for his research into coronary heart disease and arteriosclerosis, chromosomes and cancer. His wife is a physician; their only son became one too, in 1975, and Dr. Gofman continued to practice medicine during his years of nuclear research. Ironically, he was the medical director for a nuclear industrial firm during his association with Livermore, taking care of sicknesses and looking after the overall health of the firm's employees. In 1973, the $250,000 a year which his department had been receiving from the government for his cancer and chromosome work was cut off, and Dr. Gofman left the federal radiation laboratory.

Basically friendly, modest but tough, Gofman's attitude on nuclear fission power plants evolved over the years to the point where he was saying, by 1975, that "nuclear-power plants are every bit as dangerous as nuclear weapons tests." In October 1975, while delivering an address at a nuclear energy forum in San Luis Obispo, California, Dr. Gofman added a personal note, explaining why he was so concerned about the risks of low-level radiation and disease. "In my earlier medical career," Gofman said, "I used to work with cancer and leukemia patients extensively. I served as personal physician to some 30 or 40 of them in the last one to six months of their lives. It might help if every scientist and engineer had that opportunity as part of his (her) education. It is good to know what lives, and breathes, and dies behind a statistic.

"Later, I spent two years doing studies on trace elements in a variety of mentally retarded children at Sonoma State Hospital. I had a couple of days a week in the wards, seeing the human results of genetic damage. These children didn't look at all like statistics."

Gofman went on to say that a top man at the Lawrence Livermore Laboratory, whom Gofman described as a "fine person and a first-class scientist" who had "experienced intense pressure from the AEC" because of the Tamplin-Gofman findings, asked Gofman why he thought "32,000 extra deaths per year from cancer" would be "too many."

Gofman, remarking that this kind of question came from the realm of "technology without a human face," replied, "The reason is very simple. If I find myself thinking that 32,000 cancer deaths per year is *not* too many, I'll dust off my medical diploma, take it back to the dean of the medical school where I graduated, hand the diploma to the dean, and say, 'I don't deserve this diploma.'"

The same year that the Lawrence Livermore Laboratory released its findings about low-level radiation, David Lilienthal, the first chairman of the AEC, went on record to express his latent misgivings about nuclear fission power, saying: "Once a bright hope, shared by all mankind, including myself, the rash proliferation of atomic-power plants has become one of the ugliest clouds overhanging America."

A prestigious group called the Committee for Nuclear Responsibility, with 1000 supporters and Dr. Gofman as its chief spokesman, sprang up on the West Coast to warn the public about the dangers of nuclear power. By 1976 its board members included Nobel laureates Linus Pauling, chemistry; Harold Urey, chemistry; James D. Watson, biology; and biochemist George Wald, medicine. Among the other members of the board were David R. Inglis, a former senior physicist at the AEC's Argonne National Laboratory in

Illinois and a member of the team that developed the first atom bomb, Paul R. Ehrlich, a biologist from Stanford University, and author Lewis Mumford.

There were 16 nuclear plants licensed to operate in the United States in 1969, 54 were under construction, and 35 more had been ordered. Grass-roots citizen groups began to spring up in the neighborhood of reactor plants to fight nuclear power. It was also the year that Friends of the Earth was launched by editor-environmentalist David R. Brower. Brower was the executive director of the Sierra Club; he wanted the 200,000-member organization to launch a nationwide crusade against nuclear power. They had a falling out, in large part over the nuclear issue according to Brower, who left to form his new group with an overall interest in ecology—protecting wilderness areas and saving endangered species such as the whale—and a special mandate to fight nuclear power, which Brower calls the "ultimate threat" to the environment. Under Brower, who combines the expansive optimism of the West with a persuasive charm and urbanity that reflects the East, the San Francisco-based Friends of the Earth had an effective well-organized staff and over 20,000 members by 1975, affiliated Friends of the Earth organizations around the world, and a strong East Coast lobbying office in Washington, D.C. And by 1973 the prestigious Sierra Club, following ex-director Brower's lead, was also announcing that it was opposed to the building of any more reactor plants until the controversy over their dangers could be resolved.

On January 1, 1970, the National Environmental Policy Act became law. It required the AEC to prepare an environmental impact statement on reactor plants before they were built, and to weigh their benefits versus risks. But the AEC, for all practical purposes, was inclined at first to ignore this law. The bill also created a new Environmental Protection Agency, which was given the responsibility of establishing pollution and radiation standards for the nation. And 1970 was the year when the public learned, for the first time, that the AEC's nuclear waste storage tanks had been leaking at Hanford, Washington, and that some of the leaks stemmed back to the 1950s. It was also the year when two young men, self-styled philosopher-activist David Dinsmore Comey and lawyer Myron (Mike) Cherry, began to make big antinuclear noises in the Midwest that were to reverberate around the United States, and across the seas as well, where many people pattern technological programs on developments in the United States.

Comey and Cherry were particularly effective at this stage of the nuclear-power controversy because they were able to call attention to safety defects at nuclear plants; in the process, they were successful in their demands that expensive efforts be made to correct them, which helped to add millions of dollars to the capital costs of nuclear plants in America. Comey and Cherry, in short, were able to demonstrate that nuclear power was going to be very expensive because it was so very dangerous.

Comey came to Chicago in May 1970 to direct the environmental activities for the Business and Professional People for the Public Interest legal research group. Prior to that, Comey was a research associate at Cornell University's Center for International Studies and the director of the Research Institute on Soviet Affairs. While he was at Cornell Comey helped lead a successful fight in 1968 against the construction of a nuclear plant on Lake Cayuga, near Ithaca, New York.

Comey, a tenacious man, was in his early thirties, with thinning dark hair, a scholar's passion for research and analysis, an ability to reach brilliant conclusions, and a devastating wit. He was an Exeter and Princeton man, and looked like a stocky blue blood in establishment clothes who moved and acted like a free-wheeling banker. People who were fond of Comey said he was half bulldog and half pixy, and affectionately called him a snob. People in the nuclear establishment, according to the respected trade journal *Nuclear Industry,* came to regard Comey as "the most formidable—because perhaps the most rational or at least the most sharply focusing—foe of nuclear power."

Comey's method was to pore over nuclear reports prepared by the government and industry and then hang the nuclear establishment with its own material. And he soon developed a formidable expertise on reactors, how they work, and their dangers. In 1974 the U.S. Environmental Protection Agency gave Comey its First Annual Environmental Quality Award "for services that have immeasurably improved the design and safety review of nuclear reactors."

Mike Cherry, in turn, soon developed a reputation for being one of the best, most innovative antinuclear lawyers in the United States. Certainly he was the most colorful, with a calculated sense of the flamboyant. Like most successful trial lawyers, this bushy-haired product of Chicago's tough West Side was a scrapper, with a highly developed sense of split-second theater and a pronounced instinct for the jugular.

Cherry was a bright young man in one of Chicago's largest, most prestigious law firms when he got caught up in the nuclear controversy. Extremely competitive, impulsively generous, proud of being a Jew, he had been first in his class at Northwestern University's law school, and had subsequently built a reputation for being capable of playing all night and working like a demon all day.

Comey and the 30-year-old Cherry met a week after Comey arrived in Chicago. As Cherry says, "We clicked," and in June 1970 they were on their way together to fight Consumers Power Company's Palisades nuclear-power plant at South Haven, on the eastern shore of Lake Michigan, almost directly across from Chicago.

Consumers' 820-megawatt plant was all built. All it needed was an operating license from the AEC to begin generating electricity. Comey and Cherry prevented the Palisades plant from loading its fuel core for ten months and added millions of dollars in costs to the plant in an effort to improve its safety. Comey, who had learned about the dangers of thermal pollution during the fight over the nuclear plant in Ithaca, represented BPI's concern that hot water from the Palisades plant, and others like it, would harm the ecology of Lake Michigan. Cherry, on behalf of his law firm in Chicago, represented fishermen in Michigan who were afraid that hot water from the plant would be poured back into the lake to kill fish and ruin their livelihoods.

There was also a personal touch behind Cherry's involvement in Palisades. "When Consumers Power was building the plant, they sunk old navy PT boats in the lake to form a breakwater for the pipes they were putting in," Cherry recalled. "And Consumers forgot to bring 'em up. The boats broke apart, and one of the pieces went through a boat owned by one of the senior partners in my law firm. When he went to Consumers for an explanation, he said they tried to brush him off, giving him some guff like 'This is a secret government project—get the hell out of here.' So he came to me and said, 'What do you know about nuclear power?' Nothing, I said. So he told me, 'Here's the transcript, and here's the AEC law. In four or five weeks there's going to be a hearing, and I want you to go over there and give those sons of bitches some trouble, and help out the fishermen.'"

Cherry and Comey quickly developed into a terrific research-trial team, and were soon able to capture widespread attention for their well-publicized antics —including antinuclear power marches and candlelight vigils—because each, in his own way, had a knack of getting the media involved, and on his side in the Palisades case.

"We didn't know much about nuclear safety when the AEC hearing began," Cherry said, "but Comey had this sophisticated ignorance—he could 'smell' all the vulnerable points, and he had this fantastic ability to apply his studious analysis to all the documents the utility had prepared and the AEC inspection reports. And every night David would prepare my cross-examination for the next day. It soon reached the point that whenever the AEC hearing examiner wanted to know what was in the utility's application for the permit, he would ask Comey, because David knew more about it than anyone else. And I took my ten years of experience as a trial lawyer and turned this AEC hearing into a trial rather than a hearing. I also began to focus on legal issues, something the AEC, in its promotional zeal, had conveniently forgotten about in the hearing process."

Cherry opened up by charging that the Palisades hearing couldn't proceed because the AEC had failed to prepare an environmental impact statement on the Palisades plant as required by the National Environmental Policy Act of 1970. Cherry's argument went up to the U.S. Court of Appeals which ruled against him, but warned the AEC that it would have to weigh the issue of thermal pollution and other environmental aspects before the agency could issue Consumers Power a final license to operate the power plant.

With Comey doing the crucial technical research work, Cherry began to grill witnesses on what Cherry alleged were "inconsistent statements" in Consumers' application for a construction permit and then an operating license. The cross-examination went on for months. Cherry smiles as he recalls how he and Comey ran their two-man operation at the hearing behind a 20-foot-long table covered with thick manuals containing all the pertinent documents on the Palisades plant. "Comey was poised on a chair with rollers," Cherry said. "Whenever a technical question was raised, Comey would zip along behind the table on his chair —like he was on roller skates—to grab the right text. Comey's speed, instincts, and sheer ability were phenomenal."

"My theory," Cherry explained, "was that the AEC couldn't issue a license unless the applicant proved that it had actually built the plant the way it said it was going to. And then Comey and I proceeded to prove that Consumers Power had failed to build it just that way. And we predicted the accidents that would happen, which did, when the reactor began to operate. We also made every effort to insist on the use of plain English in the hearing—to stop the power company and the AEC from talking and testifying in jargon." Cherry claimed that Consumers went through five sets of lawyers during the proceedings. "First there was their house lawyer," Cherry said. "He almost had a heart attack. He didn't like yelling, so I yelled. And so on."

The big issues in the Palisades hearing were thermal pollution and radioactive discharges. By the time Consumers Power received an operating license in April 1971 to load its fuel core, Comey and Cherry forced the utility to sign a settlement, in which Consumers agreed to build closed-circuit cooling towers, at a cost of $15 million, in order to eliminate the plant's discharges of heated water into Lake Michigan. The utility also agreed to install a $5 million system to reduce radioactive discharges from the plant to a point 1 percent as high as the AEC standards would have permitted.

All told, the Palisades hearings probably cost Consumers Power an extra $50 million—for these additional safety and pollution control devices, plus the added expense of having an idle power plant on its hands for months. More important, because of the Palisades hearing, the EPA began to require many nuclear-power plants across the country to include closed-circuit cooling towers in their designs to prevent thermal pollution. And the AEC began to require better systems to hold down the discharge of radioactivity from nuclear plants. As a result, nuclear plants began to cost around $25 million to $50 million more per reactor—but they were safer. And plans to ring Lake Michigan with 20 or more nuclear-power plants began to fade because of the Comey-Cherry fight against thermal pollution.

Something kept popping up during the Palisades hearing, Cherry said, that would later dominate the issue of reactor safety systems. "Some of the AEC staff knew there were problems," Cherry said, "and every once in a while they would say to Comey and me, 'When are you going to get to the ECCS [emergency core cooling system] problem?' But we never got to it, because we never really got beyond the point of how the Palisades plant was constructed."

After the Palisades hearing was over the Chicago team of Comey and Cherry was asked by a local citizens' group led by a courageous woman named Mary Sinclair to contest another Consumers Power reactor plant, at Midland, Michigan, where the two began to raise other issues that have since become part of the lexicon of nuclear-power controversy— among them the siting of nuclear plants and their costs and reliability. By this time Cherry was convinced that the real adversary of the nuclear critics was not the nuclear utilities but the AEC—and that it was "impossible," in Cherry's words, "to get a fair hearing before the AEC."

Cherry and his Chicago law firm had also parted company by then. Cherry decided he wanted to devote most of his time to fighting nuclear power— motivated in part, Cherry admits, by the David versus Goliath syndrome: "I got a kick out of going after the big boys," he said. "And besides, I felt I was making an important contribution to society, because even though these people from the nuclear establishment might not realize what they were actually up to, they were out to poison the whole world with nuclear power."

In 1970, while Comey and Cherry were fighting the Palisades plant, Daniel F. Ford, a young economist from Harvard, and Henry W. Kendall, a nuclear high-energy physicist and professor at MIT, joined up in the East to make the Cambridge-based Union of Concerned Scientists one of the most informed and formidable groups in the nuclear-power controversy.

In common with Cherry and Comey, Ford and Kendall, in their own way, seemed like an odd, albeit highly effective pair. Ford was intense, competitive, tall and lanky, with a razor-sharp mind, and a caustic touch. Kendall, a longtime friend of David Brower's at the Sierra Club, was genial, rugged, well-heeled, and handsome; his hobbies were mountain climbing and scuba diving.

Ford's interest in nuclear power began while he was studying its financial structure. Some of the AEC documents Ford was looking at disturbed him. He took his questions to Dr. Kendall. Together they began to visit AEC laboratories and nuclear plants. Like Comey in the Midwest, Ford in the East soon developed into one of the most knowledgeable technical experts on nuclear fission power plants in the United States. He literally knew more about reactors, and the AEC programs, according to a number of scientists, than some of the people who were running them.

In 1970, the six miniscale ECCS tests in Idaho sponsored by the AEC failed. The results of the tests were published in 1971, but withheld from the public. Someone in the AEC leaked the news to Mike Cherry. He announced the failure of the tests at an AEC prehearing conference in New York, saying, "Look, this whole agency [the AEC] is dishonest. You see how they're trying to hide the whole thing."

David Comey got hold of the ECCS test documents, examined them thoroughly, and came up with some devastating conclusions that were to prove highly embarrassing to the nuclear establishment—which had been claiming, prior to the tests, that they were crucial and would vindicate the doubts about the safety of nuclear reactors. Comey turned his analytical work over to the AEC. Comey later said he felt it would be better for the "agency's image" if "it blew the whistle on itself." Instead, the AEC came out with a public statement that attempted to whitewash the failure of the ECCS tests. Comey, disappointed and annoyed, summoned reporters to his crowded cubbyhole in the BPI offices in the heart of Chicago's Loop, just up the street from the famous Picasso sculpture, where he proceeded to highlight the details of the ECCS test failures, explaining their significance; and what Comey had to report attracted a great deal of attention in this country and abroad.

Dan Ford and Henry Kendall were also picking up on the ECCS problem, and they were the ones to really develop it into a big, important issue. They began by publishing their first findings on the ECCS in the summer of 1971, then went on to raise the ECCS issue at nuclear-plant licensing hearings, charging there was a grave doubt—even within the AEC's own Division of Reactor Standards—that nuclear-plant safety systems would actually work when needed. Rather than rehash the ECCS issue at every licensing hearing, the AEC and the intervenors prepared to have a showdown on the matter.

In January 1972 the AEC began a series of public hearings on the ECCS in Bethesda, Maryland. The critics called themselves the Consolidated National Intervenors. They were largely represented by the Union of Concerned Scientists. And their technical case was mostly put together by Kendall and Ford, who was fast on his way by then to becoming a respected authority on nuclear technology. Their lawyer was Mike Cherry.

The ECCS hearings lasted 18 months. Cherry says he "viewed the ECCS hearings as the really first big chance to blast the AEC, to show, prove, how dishonest the agency really was." Ford and Kendall wanted to demonstrate—to the scientific community, government, and an increasingly aware public—that there was no guarantee that safety systems at nuclear plants would operate as touted in an emergency.

Half the time at the ECCS hearings was spent wrangling over what the AEC would open for discussion. The agency tried to limit Ford's participation on the grounds that he wasn't a physicist or a nuclear engineer—even though the proceedings, according to people who were there, revealed that Ford often knew more about reactors and their safety systems than Milton Shaw, then director of the AEC's Division of Reactor Development and Technology.

Cherry began to use the Freedom of Information Act to pry reports and data that Ford and Kendall were after from the AEC at the safeguards hearings. And dissident scientists within the agency started to leak damaging material to Ford, Kendall, and Cherry. These federal scientists and engineers alleged that Milton Shaw and other top officials in the AEC had been brushing aside problems about safety—especially those concerning the ECCS. And they also charged that Shaw had been diverting funds intended for the ECCS tests into the fast-breeder reactor development program. Then the intervenors, led by Ford, Cherry, and Kendall, handed around copies at the hearing containing information that had been leaked to them, listing over 20 points about vital functions of the ECCS that didn't work properly, or had yet to be developed.

This really created a stir: it was the first time that the regulatory and licensing divisions of the AEC had seen this list. And last there were repeated charges that AEC personnel who had misgivings about the ECCS and other aspects of reactor safety had been instructed not to disagree with official AEC policy; and that they were afraid to speak out for fear of losing their jobs.

After the ECCS hearings in Bethesda were over, a Congressional Economic Committee study on the energy outlook for the 1980s alluded to these allegations when it said, "The AEC had developed a serious credibility gap . . . by suppressing unwelcome evidence of danger and by demoting or firing researchers who have pushed their findings too vigorously. In view of the huge [federal] investment in nuclear plants, Congress might want to investigate the extent of danger in nuclear plants."

One senator, Mike Gravel, a Democrat from Alaska, had already made the dangers of nuclear power one of his major concerns. In 1971, Gravel became the first member of Congress to introduce legislation calling for a moratorium on the construction of nuclear plants; and Gravel went on to release statement after statement denouncing the dangers of nuclear power.

In late 1971 and early 1972, Mike Cherry took time out from the ECCS hearings to join lawyer Anthony Z. (Tony) Roisman of Washington, D.C., and Grand Rapids (Michigan) attorney Lewis Drain in the momentous Calvert Cliffs case. Baltimore Gas and Electric

Company had a construction permit from the AEC to build two 850-megawatt reactors at Calvert Cliffs on the Chesapeake Bay. The intervenors said the AEC had failed to prepare an environmental impact statement on the nuclear plant, as required by the National Environmental Policy Act of 1970. The AEC arbitrarily contended that the law did not apply to any utility that had applied for a permit or a license prior to March 4, 1971—16 months after the law had gone into effect.

The Calvert Cliffs plant was about half finished when the U.S. Court of Appeals in Washington, D.C. ruled in favor of the intervenors: the AEC was required, by law, to consider the ecological effects of a nuclear power plant upon the environment, and the agency had acted illegally in failing to do so. The decision was considered a landmark by the nuclear critics, who claimed it blunted the arbitrariness of the AEC; and it turned impact statements themselves into an important tool for the critics in their fight against nuclear power.

The ECCS hearings in Bethesda kept going along. People who were involved in them claim they were horribly boring and terribly important. Questions by Dan Ford and Mike Cherry brought forth disturbing revelations from the AEC and the nuclear industry like this one from L. Manning Muntzing, the AEC's director of regulation, who said, "We are not certain" that the agency's standards for the ECCS would "provide reasonable assurance that such systems will be effective in the unlikely event of a loss-of-coolant accident." Everyone involved in the nuclear controversy knew by then, of course, that the ECCS was the only barrier that stood between an overheating reactor fuel core and the public health and safety.

By the time the ECCS hearings inconclusively drew to a close in 1973 the AEC came out with an "Interim Acceptance Criteria" that permitted nuclear plants to keep operating, and new ones to be built, even though there was now substantial publicized doubt that reactor safety systems would actually work. But the

ECCS hearings did produce some concrete results. The AEC revised some of its safety standards upward, which added millions of dollars to the cost of nuclear-power plants around the United States.

The ECCS hearings also attracted the attention of consumer advocate Ralph Nader, with his ability to command attention and respect. Nader began to spend a lot of time discussing the reactor safety problem with Dan Ford and Henry Kendall, who convinced Nader that nuclear fission power was a dangerous threat. And so Nader moved into the controversy over nuclear power.

On May 31, 1973, Mike Cherry, Tony Roisman, and other attorneys filed suit in federal court on behalf of Friends of the Earth and Ralph Nader to shut down many of the nuclear-power plants in America because of the doubts about their safety. The action was directed against Dixy Lee Ray, who was head of the AEC at the time. The suit failed on legal grounds. But it did receive widespread publicity, which "we needed to wake up the public," Cherry said. "We certainly didn't plan it that way, but 'Nader versus Ray' got front-page coverage, and was on network TV. It turned out to be a way to 'merchandise' our case about the dangers of nuclear power."

Next, in May 1973, Cherry and Roisman filed the Freedom of Information Act suit against the AEC on behalf of Business and Professional People for the Public Interest and Friends of the Earth that forced the government to release the results of the updated Brookhaven report, which the AEC had buried from the public since 1965, with its figures that 45,000 people could be killed from a major nuclear-power-plant accident, while property damage could reach $17 billion.

Yet in spite of the growing opposition to nuclear power, including emerging rumblings of discontent in Congress, the multibillion-dollar nuclear establishment kept gaining ground. In this the nuclear industry was served by the Joint Committee on Atomic Energy and the AEC, which was able to generate tremendous bureaucratic momentum in behalf of nuclear power.

"The only way we can stop nuclear power," David Comey was saying in 1974, "is to torpedo it, from without, by a guerrilla attack." Comey also predicted that "the American people are going to turn against nuclear power, just like they did the war in Vietnam. They're both the same, in that they both turned into bureaucratic monsters." And 1974 was the year that Comey came out with his startling, crippling charges, presented at a federal energy hearing, that nuclear plants were unreliable, and exceedingly expensive as a result—which led to the widespread, ongoing debate about the possibility that nuclear power could fail simply because it cost too much.

In August 1974, the first draft of the Rasmussen report on the risks of nuclear accidents was released, with its comforting prediction for the nuclear establishment that the probability of a major nuclear plant accident occurring was one in a billion per reactor per year. The nuclear establishment was less pleased by another report, a long story that began on the front page of the Sunday *New York Times* of November 10, 1974. It was based on AEC documents from the past 11 years, and said the agency "repeatedly sought to suppress studies by its own scientists that found nuclear reactors were more dangerous than officially acknowledged or that raised questions about reactor safety devices."

In mid-January 1975, President Gerald Ford, responding to long-standing criticism and threatening court suits, signed a law abolishing the AEC. The new Energy Research and Development Agency was formed, which incorporated the promotional activities of the AEC; and a new Nuclear Regulatory Commission was established with the pledge to protect the public health and safety from nuclear power. Although some critics—such as old UAW man Leo Goodman —were pleased by the action, contending it vindicated all their years of accusation against the dual role of the AEC, many critics felt the NRC would be too pronuclear, and that the old nuclear agency personnel would dominate ERDA. Here the critics pointed out that almost 6000 of some 7000 employees in the new agency had transferred over from the AEC.

On April 1, 1975, a three-judge federal court in Chicago stopped the construction of an AEC-approved $400 million nuclear-power plant near Portage, Indiana, on the southern shore of Lake Michigan. In its ruling, the federal panel said the old AEC had violated its own regulations on sitings and population when it granted the builders a construction license on April 29, 1974. Noting that there were eight reactor plants within 75 miles of Chicago, with six others planned for the area, the court claimed "the AEC appears to have given no direct consideration to the clustering of nuclear-power plants around the southern end of Lake Michigan within relatively short distances from the metropolitan area of Chicago." The ruling also embraced the extremely touchy issue of the problems that might occur if people had to be evacuated from the area in the event of a nuclear-power-plant accident.

However, on November 11, 1975, the U.S. Supreme Court, in an unsigned opinion, reversed the lower court decision banning the construction of the Bailly reactor plant near the Indiana Dunes National Park. The high court said the AEC had not violated its own regulations when it gave Northern Indiana Public Service Company permission to build the atomic-power plant 1.1 miles from the city limits of Portage.

The agency's rules specify that a reactor plant must be at least 2 miles from the nearest population center. The AEC contended its regulations referred to the area where people live, not political boundaries. And the center of Portage's population is 4.5 miles from the site of the proposed reactor plant.

According to the Supreme Court, the AEC's interpretation of its own regulations was "sensible." At the same time, it ordered the federal court in Chicago to hear the remaining issues in the case, and then decide which was more important—environmental factors or the need for more electricity—in ruling whether the nuclear plant should be built or not.

Plans to evacuate people from the scene of a major

atomic accident have always been a very sore point for the nuclear utilities. Required by law, they are often painfully inept and even ludicrous, and are seldom publicized.

By the summer of 1975 Ralph Nader and public interest groups from a number of states were asking the NRC to require mass evacuation drills for people living near nuclear-power plants. And some critics were trying to force the utilities to include copies of their emergency evacuation plans with the bills they sent out. Nader said he hoped the debate over the evacuation issue would alert more and more Americans to the dangers of nuclear power.

Finally, in November 1975, with antinuclear petitions circulating in many states, Ralph Nader called another national conference in Washington, D.C. It was even more successful than the first one the year before, and revealed that the critics who were caught up in the "biggest citizens' battle of our time" were now part of a mature, sophisticated, well-organized movement from the grass roots up that had developed over the years into a powerful political, social, and moral force that felt it had caught the scent of victory.

The nuclear controversy in America had come of age.

V
Radioactive Implications

Many people in the nuclear establishment were prepared by 1975 to dismiss the dangers of low-level radiation as a relatively minor issue in the controversy over nuclear fission power. But they could be wrong; and to a staggering degree at that, because low-level radiation of a certain type might be the cause of lung cancer, most other cancers, and an important contributor to heart disease as well in America.

This radiation comes from natural sources and the nuclear fission process. Low-level radiation in nature, abetted by man's unwitting hand, could already be responsible, in whole or in part, for the majority of all nonaccidental deaths in the United States. Nuclear fission power plants and related facilities, and atom-bomb fallout, increase this burden of low-level radiation we have to contend with. And the longer nuclear power continues in this country, and the more it develops, the more we run the risk that we will fall victim to cancer and heart disease—and not only us, but generations to come.

Members of the nuclear establishment contend that the amount of radiation produced by nuclear fission power that has been, is, or will be released into the environment is so small, and its effects so minuscule, that we should be willing to regard this low-level radiation as an "acceptable risk" in return for the benefits of nuclear power. The scientist-critics of nuclear power say they're dead wrong. One of the most articulate, forceful spokesmen for the nuclear establishment is

Congressman Mike McCormack, a dynamic, wiry-haired Democrat from the state of Washington who brushes aside the allegations about the dangers of low-level radiation from nuclear fission with a tough, impatient sweep of his hand. "Most of these antinuclear people deal in gross distortions," the scientist-politician and self-proclaimed environmentalist earnestly insisted as we sat in his office on Capitol Hill in May 1975. "They deal in tiny fractions of the truth, and in many cases, in outright lies. The antinuclear movement is the McCarthyism of today in this country."

Most of the accusations about the cancerous implications of low-level radiation are directed at plutonium 239—a by-product of the nuclear fission process, and the most dangerous substance handled in quantity by man because a microscopic speck of it, inhaled into the lungs, can cause cancer. This is why plutonium has to be contained, every step of the way, from its creation to storing it as radioactive waste. It also has to be protected from terrorists and would-be bomb makers who could make a nuclear weapon with less than 20 pounds of stolen plutonium.

Most people opposed to nuclear power kept repeating the findings of a former federal expert on plutonium who said a pound of plutonium had the potential to kill 9 billion people. But by 1975 new studies were appearing to indicate that the dangers of plutonium—already thought to be horrendous—may have been grossly underestimated in the past.

Dr. Edward A. Martell, for one, came out with an evolving series of reports in 1974 and 1975 which concluded that the federal government had "minimized the effects of internal low-level alpha radiation [the same type that plutonium emits] by a factor of thousands." Martell is an environmental radiochemist and an expert on radioactive fallout from nuclear-bomb tests who lives and works in Boulder, Colorado, not far from the Rocky Flats plutonium weapons plant. He was also positive that alpha radiation causes lung cancer in cigarette smokers, and could be responsible, at least in part, for other kinds of cancer and perhaps a high percentage of deaths from heart disease as well. Martell's work with low-level alpha radiation in cigarette smokers (in particular) has been aimed at firmly establishing the cause-effect-risk relationship that exists between internal low-level alpha radiation and cancer—and heart disease, too.

This is tremendously important, in and of itself, considering the toll from tumors and cardiovascular diseases; but what Martell has really been stalking, at least in the context of this discussion, are the implications of a plutonium-fueled power economy, which could increase the risk of plutonium-emitted alpha radiation by an alarming degree. "The crucial point revolves around the relative importance of radioactivity compared to nonradioactive agents in the mutations of living cells," according to Martell. "Until this question is resolved, we have no appreciation of the full import of radioactive pollutants on human health, including genetic effects and other cell mutation-related diseases in addition to cancer."

And Dr. John Gofman, on the West Coast, was reporting that the routine leakage of plutonium in an expanding nuclear-power-plant program could, by the year 2020, increase the annual deathrate in the United States by 25 percent—that 500,000 people would die each year from cancer if we relied upon plutonium-powered plants to produce electricity. Gofman's allegation is based on a discovery he made that led him to calculate that 1 pound of reactor-grade plutonium—in the form of insoluble particles—could cause about 21 billion cases of lung cancer if the doses were averaged out between smokers and nonsmokers; and 42 billion cases of lung cancer for cigarette smokers in particular, because one of the important mechanisms of cleansing the lungs of harmful particles like plutonium is either destroyed or badly crippled by heavy smoking.

However, most scientists, especially those who do research work or act as consultants for the federal government, have consistently insisted that there is little if anything to fear from the low-level radiation produced by the nuclear fission process.

This is the position of Mike McCormack, who was an Atomic Energy Commission research scientist for 20 years before he was elected to Congress in 1970, where he became a member of the Joint Committee on Atomic Energy and chairman of the powerful subcommittee on Energy Research, Development, and Demonstration.

Back home, McCormack had been a member of the state house for four years and a state senator for ten, during which time he wrote all of Washington's energy legislation and cosponsored most of its environmental bills, including strip-mining laws, total liability for oil spills, and a Department of Ecology—which earned him the title of Mr. Environmentalist. Once he was in Congress, an appreciative power industry around the country began to call him Mr. Energy. McCormack was appointed chairman of a new Task Force on Energy —a rare honor for a freshman member of Congress; successfully fought to increase the funding for solar and geothermal energy research; supported the program to develop fusion power—which, if successful, would fuse atoms together to create energy; and was the most enthusiastic proponent of nuclear fission power in Congress.

McCormack warned me, as he does everyone, that we are running out of natural gas and oil to produce electricity, said there were a lot of pollution problems connected with mining coal and burning it, then began to read to me from one of his position papers which he had inserted into the *Congressional Record*. "Nuclear energy is the cleanest significant source of energy available with the least environmental impact of any significant option," McCormack said. "And I am certain that one of the greatest strokes of good fortune that this nation has experienced is to have our nuclear industry as well advanced as we find it today to provide much of the energy this nation will need during the next 50 years."

The nuclear establishment's goal is to build and operate between 700 and 1000 nuclear fission power plants in the United States by the year 2020. Hundreds of these would be large breeder reactors. Our large contemporary reactor plants produce 400–600 pounds of plutonium a year. A large breeder would turn out several tons of plutonium—which is virtually nonexistent in nature, although minor traces of it are found in pitchblende.

Plutonium involved in the nuclear reactor process is not only dangerous in itself. Plutonium isotopes (especially with liquid-metal fast breeders and light-water reactor units using recycled plutonium fuel elements) also produce dangerous radioactive by-products —americium and curium isotopes—which could leak or escape into the environment to enter the human food chain and build up in our liver and bones. We produced about 12,000 pounds of plutonium in this country in 1975. If we keep building atomic-power plants we could wind up putting around 400 million pounds of plutonium through the nuclear fuel cycle in the United States by the year 2000.

The nuclear establishment seems convinced it can keep most if not all of this deadly radioactive poison bottled up. But there may be accidents, theft, sabotage, or war to contend with. And even in the absence of these threats, some of this deadly radioactivity is bound to leak, according to the critics.

Dr. John Gofman says the dangers of low-level radiation stemming from the normal operation of nuclear fission power reactors and their related facilities would be so great that "the very fate of human societies may well rest upon this issue, considering the proposed handling of some 440 million pounds of plutonium—and reactor grade at that (which is over five times as hazardous, by weight, as the plutonium used in nuclear weapons)—in the next 50 years in a plutonium-based nuclear fission economy."

Dr. Donald P. Geesaman, a sensitive, sometimes caustic biophysicist, was working for the AEC's Lawrence Livermore Radiation Laboratory in California during the 1960s and early 1970s when he came out with his figures, which he kept repeating, to the dismay of the AEC, that 1 pound of plutonium could cause

9 billion cases of lung cancer. Gofman says that Gees-aman, who wound up as associate director of the University of Minneapolis's School of Public Affairs, lost his job with the AEC because of his widely publicized findings on the dangers of plutonium.

Most scientists agree that 0.001 gram of plutonium 239 is enough to kill a person within days from massive fibrosis of the lungs. And that a millionth of a gram of plutonium is far more than enough to cause cancer. Let's put this in perspective. There are 454 grams in 1 pound; plutonium has a half-life of 24,000 years. This means that plutonium will be able to trigger cancer thousands—even hundreds of thousands of years from today. The low-level radiation that plutonium gives off wears the amount of it down. If we start out with a pound of plutonium 239, we will have ½ pound left after 24,000 years, which will retain every bit of its original potency per unit of weight. Hundreds of thousands of years later there will still be a dust-sized speck of plutonium from the 1 pound we create today. And this last invisible, tasteless, and odorless particle of plutonium 239 can still cause cancer.

This is why Dr. Alvin M. Weinberg, former director of the Oak Ridge National Laboratory in Tennessee, has cautioned, "Once man has opted for nuclear power, he has committed himself to essentially perpetual surveillance of the apparatus of nuclear power."

The three kinds of radiation—alpha, beta, and gamma—produced by nuclear fission power reactors have their own characteristics. Gamma radiation is the most penetrating, and thus has commanded the most attention. We get gamma radiation from the cosmic rays of the sun, and from medical and dental X rays. The higher the dose, the more dangerous it is. The tons of thick steel and concrete that surround a nuclear reactor are there to contain these high-energy electromagnetic rays.

The beta radiation in nature comes from potassium 40, and from uranium and thorium and their radioactive descendant "daughter" products in soils, rocks, vegetation, and air. TV sets give off beta rays, and so do luminous-dial wristwatches. Tritium, a rare form of heavy hydrogen, also gives off beta rays. Beta radiation—consisting of electrons similar to those which carry electric currents, but moving much faster—can be stopped by a thin sheet of metal, thick cardboard, or a few feet of air.

Alpha radiation consists of electrically charged particles of helium gas that can be blocked by a thin piece of paper. It comes from soil, rocks, and minerals in nature that contain thorium and uranium and their radioactive daughters. Plutonium, in common with other alpha radiation emitters, can't penetrate our skin to attack the cells unless it gets in through an open wound or cut. But it can induce skin cancer if it's not washed off. And breathing specks of plutonium into the lungs can be highly dangerous, especially if the plutonium is in the form of insoluble particles, which are often difficult for the body to discharge, and hence can remain in some regions of the lung for periods of two years and more—and in the lymph nodes, liver, and bone for decades to irradiate the cells.

Even small amounts of radiation, in any form, can cause cancer and genetic defects, according to scientists. A very large high-energy dose will destroy enough cells in the body to kill anyone. Under most circumstances, the risk of disease increases in direct proportion to the amount of radiation we've been exposed to. Dr. Ralph Lapp made some calculations in the early 1970s based upon estimates that 2 million Americans would die of cancer between now and the year 2000. Radiation would be responsible for 314,000 of these deaths, according to Lapp. Of this total, Lapp predicted that 200,000 deaths would be caused by background radiation; 100,000 by medical X rays; 7000 by cosmic rays absorbed in jet airplane travel, 7000 by nuclear fallout, and only 90 by the radiation released by nuclear-power plants. Dr. Lapp, once a critic of the atomic establishment, especially over the dangers of radioactive fallout from nuclear-bomb tests, but now staunchly pronuclear, has been criticized by other scientists who believe he has underestimated the risk from fallout and reactor plants by a factor of thousands; but his assumptions at least serve to illustrate the risk relationship that exists between radiation and cancer.

The debatable point is how much radiation we are willing to accept as "safe." Obviously there's not much we can do about natural background radiation. And most people probably feel that the benefits of medical X rays far outweigh their risks. The same holds true for watching television and traveling by plane.

One thing to remember about radiation is that it tends to cumulate in our tissues. The federal standards on radiation contend that 250 rads or its equivalent in rems is a "safe" lifetime dose, while an instant dose of 400 rems would probably kill most people. But the average exposure to radiation is usually measured in much smaller doses—in thousandths of a rad or rem, called millirads or millirems. Another measure, in liquids and gas, is the curie—the amount of radiation given off by 1 gram of radium. And when we read or hear about it, it is usually in terms of microcuries or picocuries—millionths or trillionths of a curie. These tiny units of measure are a good indication of how dangerous radiation—especially from plutonium—can be.

According to the nuclear establishment, the average American received around 225 millirems of radiation a year in the early 1970s. Background radiation accounted for 110–130 millirems per year, but this could vary greatly. A person living in Denver gets about 150 millirems of cosmic radiation compared to 50 per year for a person living at sea level because people in Denver are closer to the sun. And a granite or marble building may release about 400 millirems of radioactivity a year compared to 50 from a wooden house.

The average American reportedly gets around 90 millirems a year from X rays, compared to half a millirem from nuclear-power plants. People who live right next to a reactor plant only get 5 millirems a year, according to the nuclear establishment, which is why the nuclear proponents are so fond of repeating that a person could sit on a fence next to a nuclear plant 24 hours a day, day after day, year after year, and not suffer any ill effects because of radiation.

But all these assumptions about the amount of radiation—expressed in terms of whole-body averages—which is safe or harmful may turn out to be largely meaningless when it comes to the dangers of internal low-level alpha radiation. And many scientists, including a number of Nobel prizewinners, have long argued that any amount of radiation, no matter how small, can produce harmful effects.

In 1969, the controversial Gofman-Tamplin study for the AEC reported that if everyone was exposed to 170 millirads from reactor power plants per year, which the AEC said would be harmless, it would cause thousands of cases of cancer and leukemia each year. The American Nuclear Society, however, claimed that nuclear plants would only cause one additional death from cancer every 20 years.

Dr. Linus Pauling, a world-famous chemist and Nobel laureate, employed a different methodology than Tamplin and Gofman to calculate that 170 millirems a year from nuclear plants would result in the following annual increases: 12,000 children born with gross physical or mental defects, 60,000 embryonic and neonatal deaths, 2220 leukemia cases, and 96,000 cancer deaths.

And death from cancer can be excruciatingly painful; it can also be a devastating experience for the victim's family—emotionally, physically, financially. It's our second biggest killer. Only the toll from heart disease is greater. Around 360,000 Americans died of cancer in 1975. About 370,000 are expected to die in 1976 from cancer—which will strike two out of three families in the United States, or one out of every four Americans living today, according to the American Cancer Society.

The American Cancer Society and the National Cancer Institute spent about $800 million in 1975 trying to find out what causes cancer and how to cure it. Scientists knew there was a high correlation between cigarette smoking and lung cancer; that certain viruses seemed to be present in malignant cells. And many kept saying that environmental factors were responsible for 80 percent of the cancers in the United States.

Many scientists, moreover, agreed that ionizing radiation (which tears electrons away from atoms, leaving them unstable and disruptive) could damage the chromosomes and other critical components of the cells; that radiation could alter the genes in the cells and affect heredity; that radiation could harm the body in other ways—depending on the amount of exposure, the way it enters the body, what part is irradiated, and for how long. Moreover, it doesn't matter how radiation enters the body to attack the cells. Gamma and beta radiation can penetrate our organs; alpha and beta particles and gamma rays can enter with food and drink and dust. What counts is how much radiation our cells receive. And just one malignant cell out of the billions and billions of cells in the human body can trigger a tumor. The victims of high-level radiation usually suffer injuries that can be recognized fairly soon. People exposed to low-level radiation do not. And it is harder to correlate radiation with genetic defects, because the effects may not be apparent for generations.

However, Dr. Joshua Lederberg, a Nobel laureate in genetics, says that our federal radiation standards are too lenient—that genetic defects may increase by 10 percent if we continue to adhere to them. "At least 25 percent of our health care burden is of genetic origin," Dr. Lederberg said in 1970. "This is a very conservative estimate, in view of the genetic component of such griefs as schizophrenia, diabetes, atherosclerosis, mental retardation, early senility, and many congenital malformations. In fact," Dr. Lederberg said, "as we deal with infectious disease and other environmental insults, the genetic legacy of the species will compete only with traumatic accidents as the major factor in health."

Some of the well-known diseases which are associated with heredity, such as hemophilia and sickle-cell anemia, account for only a small percentage of our fatalities. But atherosclerosis—commonly referred to as hardening of the arteries, now thought by some to be heritable, is a prime factor in some strokes and all heart attacks; and heart disease—which killed about 800,000 people in the United States in 1975—is our number one killer.

When I raised this issue of heredity with Edward Martell, he responded from Boulder, "I agree that there may be genetic as well as environmental factors which make some people more susceptible to the cardiovascular diseases and others to the incidence of cancer. However, the higher risks and earlier incidence of both coronaries and bronchial cancers among heavy smokers clearly indicate that specific agents are responsible." And Martell, as we shall see, bases his case on "the possibility of alpha radiation as the agent of atherosclerosis and cancer."

The Nuclear Regulatory Commission, which succeeded the AEC in 1975, contends that the radioactive discharges from nuclear-power plants are kept as "low as practicable." Practicable, of course, means "economically feasible." And the nuclear establishment boasts that reactor plants contain 99.99 percent or more of all the radioactive poisons they produce. But the important measure is the total amount of radioactivity released into the environment, rather than the percentage which is contained. A nuclear fuel core meltdown, of course, could release tremendous quantities of radioactive poisons.

These radioactive elements have their own characteristics. Some have extremely long radioactive lives: strontium 90 can continue to trigger cancer for hundreds of years; plutonium, of course, can do it for hundreds of thousands of years. Strontium 90 gets into the bones and teeth, cesium 137 into the muscles, krypton 85 into fatty tissues. The thyroid gland has a special affinity for radioactive iodine 131, and a child's small thyroid is much more vulnerable than an adult's.

The amount of radioactivity released by a nuclear-power plant might seem minuscule; in fact, it is often less than 1 percent of the amount allowed by law. But a cow grazing on radioactive grass can concentrate iodine 131 in its milk and pass it on to a child in the form of fresh dairy products. The same is true of strontium 89 and strontium 90. By the end of a year or more the amount of strontium 90 in a quart of milk or a pound of cheese may be thousands of times as high as it would be in the air or the soil around a nuclear-power plant.

Most reactor plants are in rural areas—where there are deer and rabbits and fowl and squirrels and other animals that are hunted in season. These creatures are not confined to pasture. They roam about, and they could be feeding on radioactive food and drinking contaminated water, which they could concentrate in their bodies and build up. What about them? And fish? Most of our nuclear plants are next to or near streams, rivers, lakes, or the ocean. I've seen people fishing next to nuclear plants. Radiation that's discharged into rivers and lakes keeps building up in the aquatic food chain—from bacteria to algae to zooplankton to fish, which can harbor radiation levels thousands of times as high as the trickle released from a nuclear plant or related facilities.

It's obvious, then, that the human food chain, in all its complex and little-known pathways, is an extremely important factor when it comes to assessing the dangers of reactor-produced radiation. The nuclear critics charge that the radiation guides laid down by the federal agencies tend to ignore the implications of these radioactive food chains.

There was relatively little concern about the dangers of low-level radiation until the atom-bomb testing years of the 1950s and early 1960s. In fact, many scientists and members of the public at large were unaware of low-level radiation until the ban-the-bomb movement in the United States and elsewhere began to publicize the health threat from radioactive fallout. As a result, many attempts to measure the risks of low-level radiation refer back to the findings from fallout.

Dr. John Gofman made use of them when he brought forth some very disturbing statistics in 1971.

The average American, according to Gofman, received about 19 millirads of radiation per year from all the fallout that was released by the atom-bomb tests through 1963. Next he cited AEC estimates on how much fission reactor energy would be produced within the next 30–50 years; and with it, the yearly amount of radioactivity that would be created by the 1000 nuclear plants the nuclear proponents hoped to build and operate by the year 2000. Then Gofman calculated the amount of radioactivity that would be released from these 1000 plants if only one one-thousandth of the fission products escaped—which, Gofman concluded, would be 200 times the annual rate of radioactivity the American public received during the bomb-testing years: 3800 millirads per year from reactor plants compared to 19 millirads per year from fallout.

These figures were anathema to the nuclear establishment. By 1974 pronuclear advocates such as Congressman Mike McCormack were repeating their version that we would get "only 0.425 millirem of radiation per year from the operation of all nuclear plants and their related activities . . . if we assume 1000 nuclear power reactors on line in the year 2000."

It has been difficult to assess the amount of radioactivity released by reactor power plants because monitoring systems have often been largely inadequate. Pennsylvania is one state that has found this out—but only because of antinuclear critic Dr. Ernest J. Sternglass, a professor of radiation physics at the University of Pittsburgh, who went around charging that infant mortality rates had risen in the vicinity of nuclear-power plants, and that the number of infants with leukemia and cancer had increased around reactors because pregnant women had been exposed to radiation—when studies have indicated that the fetus is 50 times as vulnerable to radiation as an adult.

Sternglass in action is an intense, exuberant man who wields large statistical charts and spills out his correlations in a rapid, outraged voice touched by a slight lisp. Behind the Sternglass death charts is his allegation that reactor plants release more radiation than most people suspect—often 100 to 1000 times as much, according to Sternglass. He has appeared at many licensing hearings to oppose nuclear plants, and has lectured widely on the dangers of low-level radiation. His name is a constant in the literature of the nuclear controversy. The Atomic Industrial Forum, the American Nuclear Society, the nuclear utilities—all have spent a small fortune trying to counter Sternglass's antinuclear statistics. He has been called a charlatan. The American Nuclear Society claims he cannot prove his charges, that his statistical methods are erroneous, and that his calculations have been repudiated by a number of state health departments, the Health Physics Society, and the American Academy of Pediatrics.

But his allegations seem to plant at least a few seeds of doubt whenever Sternglass appears; and there are a number of scientists who believe his work merits further consideration. At the very least, they say, society owes a debt of gratitude to Sternglass for the questions he has raised, which began with his studies on the relationship, if any, between fallout and infant mortality rates. This index was considered significant because improvements in health and medical care in the 1930s and 1940s in the United States had driven down the number of infants who died in their first year. When the atom-bomb tests began in 1951, the infant mortality rate suddenly reversed itself and began to climb, according to Sternglass and others. When the nuclear tests stopped in 1963, the rate began to drop again. Then it picked up. Sternglass's investigations led him to conclude that low-level radiation from fallout and reactor plants has been and is killing babies. Some scientists began to think his thesis had merit. Dr. David R. Inglis, for one, a former AEC senior research physicist who was a member of the team that developed the atom bomb, said in 1972 that "despite . . . reservations, the collection of cases that Sternglass presents would seem to indicate that there is a relationship between fallout and infant mortality of the general nature he suggests."

Sternglass claims he wanted the AEC to conduct studies on this relationship between radiation, disease, and death; but couldn't interest the agency in the project, which is why he began his own statistical probe.

Sternglass soon created a furor by charging that the presence of Shippingport reactor plant in Pennsylvania was responsible for apparent increases in cancer, leukemia, and infant mortality all along the Ohio River from Pittsburgh to Cincinnati. In 1972 Duquesne Light Company tried to claim another first for the nation's first atomic-power plant—that it was the "cleanest, safest nuclear plant in the world," contending it had not released a bit of radioactive gas in a year. But Sternglass claimed that strontium 90, cesium 137, and iodine 131 in the air, soil, and milk samplings around the plant were way above normal. His allegations were resoundingly attacked by the utility's consultants. But when Duquesne Light asked for a construction permit to build two large reactor units at Shippingport, the mayor of Pittsburgh, civic and environmental groups, and other anxious citizens began to raise a lot of questions and objections.

Governor Milton J. Shapp of Pennsylvania appointed a special study group to look into Sternglass's charges about Shippingport. Reporting back in July 1974, this Fact Finding Committee said there was "no substantial evidence" that radiation from Shippingport was higher than the allowable limits and "no systematic evidence" that the plant was injuring people who lived in its vicinity. But the report added there just wasn't enough information to support or refute Sternglass's allegations because radiation monitoring systems at the plant were "inadequately devised and carried out" by the AEC and Duquesne Light. The governor's special committee concluded by asking the federal government to establish "accurate and reliable" monitoring programs for nuclear facilities, and urged Pennsylvania to initiate its own health statistics program to determine if there was a relationship between nuclear-power plants and disease.

John Gofman contends that Sternglass has "made

a magnificent contribution for showing there's a great, neglected need to monitor around nuclear-power plants. The government should fund him," Gofman said, "if only for raising the kind of questions that need to be asked."

Dr. Edward Martell, an environmental radiochemist with the National Center for Atmospheric Research in Boulder, Colorado, says, "I think Sternglass is right—at least, intuitively so, even though he hasn't been too successful in gathering persuasive evidence." Martell, a solid, ruddy man with gray-white hair and an easy-going manner, is a different type of scientist from Sternglass, who seems impulsive, bold, and often in a hurry to announce his calculations. Martell, highly disciplined and respected by his peers, is methodical, careful, cautious, and given to understatement. If he were positive, Martell would say, "I suspect," and in a soft voice at that.

When I first met Edward Martell in the summer of 1975, after hearing that he was one of the leading authorities on plutonium in the country, I found this rather courtly scientist smoldering with excitement. He had just completed four intensive years of study, research, and experimentation on low-level alpha radiation. "If I'm right," Martell told me, "it means that natural low-level radiation from insoluble alpha particles in the body may be the principal agent of human cancer—especially lung cancer among cigarette smokers—as well as heart disease and strokes."

When I heard this my ears really perked up; I knew that plutonium gives off alpha radiation, and that it is especially dangerous in the form of insoluble particles in the lung. And I also knew that over 80,000 people were already dying from lung cancer each year in the United States.

"But the AEC minimized the effects of internal low-level radiation by a factor of thousands" for atomic workers and the general public, Martell continued. "If I'm right," he said, getting up, sitting down, walking around his desk, stepping briskly over to the blackboard in his office to jot down a dizzying stream of equations, explaining them, erasing them and then writing more on the board, "it is earthshaking. And the only way I could be wrong," he cautioned, "is if cancer is not due to multiple structural changes in the critical constituents in a single cell.

"If I'm right, fission reactors—especially the breeder, because of plutonium production—are totally unacceptable. Nuclear fission is already in trouble," Martell said, "because of reactor safety and plutonium safeguards. But it wouldn't require a catastrophic accident involving plutonium . . . because every little accident, in fuel transportation or processing, could gradually make this world uninhabitable."

On paper, or at an official hearing, Martell is much more reticent than he is in the privacy of his office; in public his style is low-key, suggestive, full of "may be . . . could be . . . suggests . . . implications . . ."—accompanied by pleas for more study and research into the dangers of low-level radiation, underscored by a warning that "serious uncertainties in the cancer risks attributable to internal alpha emitters must be resolved before we are irretrievably committed to a nuclear energy program." Behind Martell's public caution is his feeling that "if you publish too early, make one obvious error, no one [meaning scientists] will believe you—nor will they if you sound too sure of yourself. But I've waited until enough proof was in. I'm confident now," he told me. "You see," Martell explained, "the AEC was concerned about high-energy radiation from reactors. So they didn't pay much attention to internal low-level alpha activity from fallout and reactors, or worry about it, because it was only about equal to or even less than the natural alpha level. Besides, and this is the really important point, the AEC ignored the whole issue of relative risks for soluble versus insoluble alpha activity. And this," Martell said, "is the critical issue for cancer risks."

All natural alpha radiation emitters—such as polonium 210—which are produced by the decay of radon gas given off by radium and uranium are soluble in body fluids. We take them into our bodies every day of our lives—in our food and drinking water and when we inhale microscopic bits of dust containing alpha particles into our lungs. These soluble radioisotopes are dissolved in the lungs and evenly distributed in fluid, which moves them through our body and out of our systems. It takes about a day, more or less, to discharge these natural soluble alpha emitters from the lungs. Because there's a continual daily balance between the amount we're taking in and discharging, there's always about the same amount of radioactivity in the lungs from soluble "background" alpha emitters. Because these soluble alpha emitters are evenly spread in fluid throughout the lung, each of the tens of billions of cells in the tissues of the lung is exposed to a low uniform dosage of alpha radiation every day.

There's nothing we can do about this. We can't stop breathing, eating, or drinking. However, the risk of getting cancer from this steady flow of natural soluble alpha radioactivity in and out of our bodies is very slight—even though a small portion of radioactive lead 210 and radium 226, which are chemically like calcium, is deposited and builds up in bone—because the uniformly distributed radioactive dose is so low. But some of these dust-sized specks in nature are converted into insoluble radioactive particles—by forest fires and various industrial processes which spew insoluble alpha particles from factory stacks into the atmosphere. And by smoking cigarettes. A terribly important point, considering that one out of every two adults in America smokes cigarettes on a daily basis. Which means we're already talking about 70–80 million Americans. And every now and then, some insoluble alpha particles will be absorbed in lung tissue for periods of time up to two years or even more.

Some nuclear proponents claim that alpha particles are not especially dangerous because they discharge all their energy in such a short distance—which is why their rays can be blocked by a sheet of paper. There's another way, however, of looking at alpha rays: they attack with such dense force that they travel only a tiny distance before they exhaust all of their energy;

but before they stop, they discharge intense ionizing radiation in the tissue they encounter.

Because alpha rays have a very short range, the typical burden of insoluble alpha emitters in the lung can irradiate only about 1 gram of tissue or less. But the rays keep pounding away, slowly, repeatedly, within their short alpha range. And the cells are where the mystery and the clues of life and death reside. Therefore, in order to understand what nuclear fission reactor power is really all about, we have to tear our vision away from the giant reactor plant and move inward, toward the tiny human cell—and cancer.

Every normal human cell—except during certain stages in the male and female reproductive cells—has 46 chromosomes. And every chromosome has hundreds or thousands of genes in it—which are our units of inherited information, composed largely of deoxyribonucleic acid, or DNA, which direct the normal function and behavior of the cell. In addition there are other critical constituents that play vital roles in the function of the cell. When we're young we grow because these normal cells keep splitting to recreate themselves in a stupendous, orderly, healthy profusion. And when our tissues are invaded and the cells are killed by anything from a cut to disease to radiation, the uninjured cells in the body can split in two to reproduce healthy normal replicas of themselves to replace their dead or missing neighbors. Cells that are injured, but not too badly, can still carry on—perhaps not as well as before, but they can still function. And life goes on.

However, if too many cells are mutated or killed, the organ or part of the body where the tissue was damaged can fail to perform normally, or die—and this applies to the whole organism too, depending on the extent of cell damage. And we begin to age, and die, when our cells are heavily damaged, or die faster than the remaining healthy cells can multiply to replace them.

The cells in normal, healthy tissue are a beautiful, wondrous sight to behold under the microscope. They are simply breathtaking in their clarity and orderliness.

In contrast, the tissue in a cancerous tumor is raw, ugly, frightening—a tortuous jungle, a swamp without rhyme or reason, crawling with broken chromosomes and hideous-looking mutated cells.

When an alpha ray from an insoluble particle attacks a cell, or cells, it rips through like a cannonball. Most of these direct hits will kill the cell, and a dead cell cannot reproduce itself—healthy or cancerous. (This is the principle behind the use of radiation therapy: to kill off malignant cells to stop them from spreading.) But a glancing blow or a near-miss from an alpha ray can cause the kind of irreversible structural breaks that are most likely to produce a malignant cell, according to Martell.

Moreover, Martell has pointed out that the insoluble alpha-emitting particles have a tendency to cluster together in the lung; therefore, a single gram of tissue containing an insoluble particle burden in the lungs of cigarette smokers will be exposed to from 100 to 10,000 times as much radiation as would be received from natural soluble alpha activity moving through the lung in body fluid, when each of the 570 bloodless grams in the lung would get the same small uniform dose of natural radiation—too small to cause much harm, except in special cases, when uranium miners, for example, are exposed to large, cumulative amounts of soluble alpha radioactivity.

Another way of looking at this revolutionary phenomenon, according to Martell, is this: 1 gram of tissue in smokers gets 1000 times as much alpha radiation from embedded insoluble smoke particles as the whole lung would get from natural alpha activity, which means that the risk of getting a tumor from an insoluble particle is 1000 times as high as it would be from soluble alpha activity.

The federal radiation guides assume that alpha particles expose the whole lung—and each of the 50 billion cells in it—to the same low average amount of radiation, which, scientists like Martell and John Gofman contend, unjustifiably assumes that soluble and insoluble alpha particles have the same effect. This could have tragic implications when it comes to calculating the cancer risks of plutonium—because the relevant index is not the measure of radiation in the whole lung, according to Martell, Gofman, and others, but the amount of alpha radiation in less than 1 gram of tissue. Moreover, the local alpha radiation that a particle of plutonium gives off is billions of times as high as the amount emitted by dissolved natural alpha activity from an element like polonium 210. A thousandth of a gram of plutonium can be immediately fatal because it kills so many cells. Federal radiation guides, however, do not take into account the cumulative dangers from doses released by insoluble alpha particles that remain in the lung for periods of time up to two years and even more. This is why some scientist-critics point out that the nuclear proponents refuse to recognize that there's likely to be a greater risk from nonuniform than uniform distribution of plutonium—or even make statements to the contrary. John W. Simpson, for example, the director of Westinghouse's reactor division and a past president of the American Nuclear Society, was president of the Atomic Industrial Forum in 1975 when he said: "Ingested plutonium in the insoluble oxide form [the one with which we are primarily concerned] is not nearly as dangerous as in the soluble form."

Martel, Gofman, and other scientists strenuously disagree with this contention. There's no doubt that it is insoluble plutonium that the general public is exposed to, and has to worry about. Plutonium readily combines with water vapor in air to form insoluble plutonium oxide particles. When plutonium metal corrodes, it forms insoluble dioxide particles. And when plutonium burns—which it does, unpredictably and spontaneously—it forms insoluble particles of plutonium oxide that are small enough to inhale. What this means, according to the scientist-critics of nuclear power, is that much of the plutonium that escapes into the environment is going to wind up in the form of insoluble particles.

About 5 tons of insoluble particles of plutonium dioxide were released to the atmosphere in fallout during the atom-bomb testing years. And additional tons are trapped at underground nuclear test sites. A major nuclear-plant accident—especially at a breeder reactor facility—could release an awesome amount of plutonium that would disperse as insoluble particles. And people who work with plutonium in weapons plants and fuel reprocessing and fabrication facilities have to be extremely careful—because plutonium is what the scientists call a very refractory or obstinate element: it keeps bursting into flame, for example, when it comes into contact with moist air.

Handling plutonium can be fraught with risk. The federal radiation standards governing the amount of plutonium the authorities think we can inhale without harm is measured in picocuries—trillionths of a curie. But a number of nuclear critics think the current standards have "underestimated the dangers of low-level radiation by a factor of thousands," in the words of Martell, because they have failed to distinguish between the cancer risks from soluble compounds, and insoluble particles that emit alpha rays. "AEC experts and other specialists have overlooked this complicated possibility," Martell told me. And his story, which follows, and what he has found out about low-level radiation, could have a profound meaning for all of us.

When I met Edward Martell in 1975, he had been with the National Center for Atmospheric Research for 13 years, where he won attention and respect for his work in atmospheric chemistry, air pollution, and radioactive aerosol research. Before moving to Colorado, the West Point graduate and former army lieutenant colonel was with the Armed Forces Special Weapons Project studying the effects of radioactive fallout from 1950 to 1954. Martell then put in two postdoctoral years at the Enrico Fermi Institute for Nuclear Studies, and went on to head a radioactive fallout research group at the Air Force Cambridge Research Center. It was about this time that he made himself unpopular with the AEC by disclosing that radioactive iodine 131 was leaking into the atmosphere from underground nuclear-bomb test sites in Nevada.

The National Center for Atmospheric Research,

funded by the National Science Foundation, sits on a lofty mesa in the southwest corner of Boulder, where the Great Plains end and the Rocky Mountains begin. It's a beautiful, creative, vital setting where the wind comes whipping down over the mountains with sudden, breathtaking force to sweep east over the dry plains toward nearby Denver and beyond. The Rocky Flats plutonium weapons and reprocessing plant is 6 miles south of the Center for Atmospheric Research. Martell could look out of the windows at the Center and see the plutonium plant on a clear day. But he never gave it much thought before 1969.

The $200 million Rocky Flats facility began producing plutonium triggers for nuclear warheads in 1953. The plant, owned by the AEC, was operated under contract by Dow Chemical Company until the summer of 1975, when Rockwell International, a subsidiary of Atomics International, was awarded a $70 million federal contract to run the facility. Rocky Flats has an important economic impact on the area. Its work force, which reached a high of 3700 in 1971, was about 2800 when I visited the plant in 1975.

The plant was relatively isolated when it was built in the early 1950s. But the heavily populated suburbs west of Denver have reached to within 6 miles of Rocky Flats, and downtown Denver is only about 16 miles from the nuclear weapons plant, which sits in the middle of 6500 acres of flat, arid land and sagebrush.

The metallic plutonium at Rocky Flats is refined and precisely shaped for warheads. Most of the work goes on inside large enclosed glass cubes called dry boxes, or glove boxes, because of the heavy lead-lined gloves which protrude inward from the glass walls, into which workers put their hands to manipulate the plutonium inside the enclosed rooms.

There were a couple of hundred fires and more cases of radiation contamination at Rocky Flats during the 22 years Dow Chemical was operating the plant, according to federal figures. And thousands of 55-gallon drums containing plutonium-contaminated cutting oil that had been used to machine plutonium—which were stored out in the open—corroded and leaked over a period of years. On September 11, 1957, there was a $1 million fire at Rocky Flats which was touched off by the spontaneous combustion of plutonium in a dry box. It spread through a ventilation system and released a plume of smoke containing traces of plutonium, according to reports of the fire. Then a worker lost a finger in an explosion on June 12, 1964. The blast occurred when he inadvertently dropped some plutonium chips into a bucket of carbon tetrachloride. And 400 employees were exposed to plutonium oxide fumes when plutonium chips caught fire in a large room on October 15, 1965.

The local nuclear critics say all this was only a warm-up. On May 11, 1969, at 2:27 p.m., a fire broke out in one of the buildings at Rocky Flats. Some observers reported seeing dense black fumes for a short period, then light gray smoke for quite some time. Some people at the plant, however, insist that there was no smoke outside the burning building.

At first the AEC said it was a $2 million fire. Then it turned out to be a $50 million blaze, one of the most costly industrial fires in American history. Flames damaged a portion of the plant and put it out of operation for months. Twenty million dollars' worth of plutonium was burned. Some of the firemen who fought the blaze were contaminated, and Dow Chemical hired several hundred temporary employees, including students, to clean up after the fire. The Rocky Flats people I talked to claim no smoke escaped; that "high efficiency air filters contained 99.97 percent of the plutonium particles from the fire" and no one was hurt in the fire or contaminated by plutonium.

A group called the Colorado Committee for Environmental Information started to investigate the plutonium fire. It was a small group, with members from the University of Colorado, the National Oceanic and Atmospheric Administration, and the National Center for Atmospheric Research. Edward Martell headed the group's subcommittee that had been formed to investigate the fire and its implications. They met with representatives of the AEC and Dow Chemical Company to discuss these issues; the nuclear critics also began to measure plutonium levels in soils in the vicinity of Rocky Flats. "That's when I began to realize how little the AEC knew about the health hazards of plutonium," Martell said.

In a report published in May 1970, the nuclear critics alleged there was "100 to 1000 times as much plutonium in the local environment as there would be if good containment practices were continually maintained at the Rocky Flats plant." And the ecologists claimed that most of the plutonium deposited on soil downwind of the plant was in the form of insoluble particles of plutonium, of which many were of respirable size. Most of these particles, according to the nuclear critics, were on or near the surface of the soil, which meant that the strong winds—at 100 miles an hour or more—from the Rockies that often buffet the foothills would probably keep stirring these particles of plutonium up and into the air.

The nuclear critics began to wonder, out loud and with considerable publicity, how many of these insoluble plutonium particles would wind up in the lungs of people living around the plant and in nearby Denver —and how many of these particles a person could inhale without running the risk of cancer.

Dow Chemical, which had been monitoring for radiation around the plant, but not for plutonium in particular, began to measure the discharge of its primary product. The Colorado Department of Health stepped up its monitoring program around Rocky Flats. And the AEC, responding to the Colorado Committee for Environmental Information, restated its "conviction that such trace amounts present no risks to the health of employees in the plant or to citizens in the surrounding area."

The plutonium level, according to the AEC, was "minute"—far below the level established by federal radiation guides as dangerous. "Air samples in the vicinity" after the plutonium-contaminated oil spill of 1968 "showed higher radiation levels than normal," the AEC said, "but still well below applicable standards."

The nuclear critics estimated that tens to hundreds of grams of plutonium escaped from Rocky Flats. The AEC said, "Most likely less than an ounce, spread over 50 miles" had gotten out, of which ⅕ ounce had been "released from the plant's ventilator system under carefully controlled conditions during normal operations . . . since 1953." The agency also said that most of the plutonium from the 1969 fire fell on the roofs of buildings at the plant, where it was cleaned up.

By this time, because of his concern about plutonium, Martell was reading every bit of information he could lay his hands on about cancer risks from low-level alpha radiation. And Martell and Stewart E. Poet, his research associate at the atmospheric center, finally agreed with the AEC and Dow Chemical that most of the offsite plutonium around the plant had escaped from the oil-spill area prior to the fire in 1969. While Martell and Poet were monitoring in the vicinity of Rocky Flats, they made another important discovery. There was radioactive americium 241 around the plant. This is a dangerous isotope (because it moves so readily into the human food chain) with a 460-year half-life. It is the by-product of plutonium 241, an isotope with a 14-year half-life. And the two scientists said the plutonium in the vicinity of Rocky Flats was still 10 to 100 times as high as the amount of plutonium that had been deposited in the area by fallout from nuclear weapons tests in the atmosphere.

Edward Martell, as he later wrote in 1971, had this important point firmly in mind from the beginning of his extensive research, study, and laboratory work on internal low-level alpha radiation and cancer: "When plutonium is burned in air, the particles produced are highly insoluble plutonium dioxide particles of respirable size and high specific alpha activity." He warned in 1971 that "inhaled particles of this type carry carcinogenic risks of uncertain magnitude." However, in April 1975, a report published by the

Rocky Flats plant continued to maintain that all radio-active particles released by the plant "are assumed to be soluble for purposes of comparison with appropriate concentration standards"; and that the total dose was less than one-tenth of the radiation background level in Colorado.

By 1972 Martell was saying, "If the AEC is allowed to pursue a plutonium fast-breeder program, we must first obtain an adequately comprehensive evaluation of the chronic effects of low-level plutonium on man." In 1975 he was still insisting that "the present knowledge [about plutonium] is fragmentary and based on less than three decades of experience with fallout and nuclear accidents," and said what we need to understand is how these radioactive particles might be redistributed in the environment for thousands of years.

In his quest to find out everything he possibly could about plutonium and lung cancer, Martell started out by studying the famous 1964 report of the U.S. surgeon general on the dangers of cigarette smoking and cancer. It was the product of the biggest epidemiological study in history; its pages were loaded with fascinating information—not only about cancer, but the correlations between smoking and every other disease as well, broken down by age, sex, occupation, and place.

Martell was aware that two radiobiologists at the Harvard University School of Public Health had suggested in 1964 that smokers could be getting a cancerous dose of alpha radiation in their lungs from inhaled polonium 210. The two scientists were Edward P. Radford and Vilma Hunt. Martell got in touch with Dr. Radford and he has been working closely ever since with him. Dr. Radford is now at the Johns Hopkins University School of Hygiene and Public Health in Baltimore, where he has been studying the effects of radiation on lung tissue. In the 1960s, scientists knew there was more polonium alpha radio-activity in the lungs of smokers than of nonsmokers. But there was only a tiny bit more in the lungs of smokers—a few picocuries. This was a puzzle. Scientists knew that natural polonium 210 was soluble. Therefore, they assumed that alpha activity drawn into the lungs with cigarette smoke particles would be washed out of the lungs because the particles were soluble. Nevertheless, spots in the lung, confined to a gram of tissue, contained high concentrations of polonium 210. The scientists couldn't figure out how this polonium 210 was accumulating in the bronchus—the area where most smokers' lung tumors start. But most researchers didn't think and many say they still don't believe the matter was worth pursuing because the amount of alpha radiation in the lung as a whole seemed very low for both smokers and nonsmokers who have died of lung cancer.

This was Martell's first important clue: man-made plutonium 239 gives off ionizing alpha rays. And the excess of natural polonium 210 nestled in the tissues of smokers was an alpha emitter too. Obviously something was happening here, and Martell was determined to find out what it was, if he could. And this is what he discovered: When radioactive radon gas (from uranium and radium in the natural environment) decays the gas produces three radioactive elements—bismuth, which is short-lived, lead, and polonium—which collect on tiny particles of atmospheric dust less than 0.1 micron in diameter. Some of these particles gather on tobacco leaves. And what follows is really "hairy."

There are tiny hairs on the surfaces of the tobacco leaf called trichomes, which have sticky glandular heads on their tips that collect radioactive dust particles. Some of these particles that stick to the tobacco leaf hairs originate in the natural soil while others arrive on the wind. And tobacco soil is treated with fertilizers that have heavy concentrations of phosphate which contain rich amounts of natural radium. So the tips of the tiny hairs on the tobacco leaf surfaces are usually loaded with radioactive particles. Also, the smaller the particle, the higher the concentration of radioactivity. The lead 210 content of trichome ash is as radioactive as pure uranium or thorium.

When a cigarette is lit and the smoker drags on it, the heat at the tip of the burning cigarette reaches temperatures between 750° and close to 1000° C. This volatizes the polonium 210 and fuses the lead 210-coated hairs into insoluble particles that are drawn into the lungs with cigarette smoke. The volatile polonium 210 that gets sucked into the lungs from the cigarette gets washed out quickly because it's soluble, whereas the particles of lead 210 can become trapped in tissue because they're insoluble.

Lead 210 has a radioactive half-life of 22 years. The beta rays the lead 210 gives off are too weak to do much damage to the cells in the lung. But when lead 210 gives off radiation it turns first into bismuth 210 and then into polonium 210, which has a half-life of 138 days. This means that as long as there's lead 210 in the lungs there will be polonium 210 too; and the dense column of ionizing alpha radiation emitted by polonium 210 is strong enough to cause plenty of damage to the cells it encounters. Most of the lead 210 that's inhaled into the lungs from cigarettes will be discharged from the body, but some of these insoluble particles will persist in the bronchi and elsewhere in the lungs. In takes anywhere from months to years for the body to dislodge some of these insoluble radioactive particles embedded in lung tissue.

During the course of a lifetime we inhale hundreds of grams of insoluble dust particles. Not all of them, of course, are radioactive. When we reach the age of 70, most of us will have about 1½ grams of insoluble dust particles in our lungs. A young adult may only have a fraction of that amount, but men who work in mines, or asbestos plants, for example, where there's plenty of dust, will collect a lot more insoluble dust in their lungs than the average person. This dust, according to Martell, builds up, clogs the smaller lung passages, and obstructs the flow of mucus fluid, which decreases its ability to cleanse the lungs. This in turn increases the likelihood that insoluble radioactive particles—from cigarettes, fallout, nuclear facilities, industrial pollutants, and other sources of insoluble alpha radiation emitters—will be able to lodge and collect in human tissue.

Cigarettes, in particular, seem to be an especially heavy source of insoluble radioactive particles because tobacco fields are fertilized with radium-rich "superphosphate of lime"—a mixture of calcium acid phosphates and gypsum—or pure calcium acid phosphate called triple superphosphate.

When calcium acid phosphate is heated in a burning cigarette, it converts into insoluble calcium metaphosphate, "a compound that fuses at a temperature of 975° C, just below the maximum temperature attainable behind the burning cone of a cigarette," according to Martell. This doesn't happen with ammoniated phosphates. Which leads Martell to suggest that using ammonium phosphate fertilizers would not only greatly reduce the radium contamination but also might eliminate, or at least reduce, the number of insoluble smoke particles and cut down their size. Therefore, it should be possible to produce and market a cigarette with a much lower cancer risk.

Meanwhile, the insoluble alpha-emitting particles that are trapped in tissue tend to cluster together to transmit a high accumulated dose of alpha radiation to the tens of millions of cells in 1 gram of lung tissue. A smoker will get about 200 rems in a 25-year period, according to Martell—more than enough to cause cancer. A natural alpha-emitting radiation particle that's trapped in bronchial tissue remains there for six months or more before it's discharged from the bronchial tumor sites, according to Martell. Therefore, if a cigarette smoker gives up smoking, the bronchi will clear of high localized levels of radioactive alpha activity within two years or so, which is why the risk of bronchial cancer drops off for the cigarette smoker who quits, according to Martell.

After four intense years of study, research, observation, and laboratory experiments with insoluble alpha particles, Martell was ready to begin airing his findings in 1974. Characteristically, his first summations were cautious and tentative. "Alpha radiation from polonium 210 may be the primary agent of bronchial cancers in smokers," he wrote in the May 17, 1974, issue of *Nature,* pointing out that even though the level of radiation in the lungs as a whole was increased by only a few picocuries because of insoluble alpha particles, it was concentrated in a small volume of tissue.

By 1975 the reports that Martell was publishing were more conclusive. He was saying that the insoluble alpha particles were 1000 times as dangerous as the same amounts of soluble alpha activity. He also said alpha radiation was probably more of a cancer threat to most of us than gamma and beta radiation. An alpha ray, which travels only a short distance, produces a dense column of ions to strike with the impact of a dumdum bullet. Most direct hits in the cell will kill it. But glancing blows will shatter the cell's critical constituents with a number of close, complex breaks which make it impossible for most of them to regroup in normal healthy fashion; instead, the broken constituents form a number of abnormal recombinations which we call mutated cells. The effects of low doses of gamma rays and X rays are much different in that their ions are spread out and diffused. If we liken an alpha track to a single blunt bullet, X rays and gamma rays are more like scattered buckshot, moving through tissue to break a critical cell component here and there with widely spaced individual breaks that give these split components a good chance to find each other, regroup, and carry on again in normal healthy cell fashion.

Martell also suspects that radiation may be a contributing factor to the process of aging—which occurs when the number of mutated cells in the body increases and can no longer be replaced by normal cells. Conversely, the cells of a growing child keep proliferating. And it's a "well-established fact," in Martell's words, that tissues with a lot of rapidly dividing cells "suffer the earliest and most severe radiation damage"—which is why the fetus and the youngster are more susceptible to radiation damage than adults. It's worth noting, in this context, that cancer kills more children aged 3 to 14 than any other disease in America, according to the American Cancer Society.

Martell, in his work, pointed out that current studies by Dr. Albert Leven of Lund University's Institute of Genetics in Sweden, one of the leading researchers on chromosome structure in human tumors, and other scientists, indicate that cancer is the end product of complicated processes which give rise to a number of chromosome structural changes. "And alpha-emitting particles produce a lot of mutations," is the way Martell puts it. "Continuously, the mutated cells that have the highest mitotic rate take over: they can outgrow the healthy cells and other mutated cells."

Tumors, incidentally, contain a maze of fantastically complex structural breaks and an amazing variety of mutated cells—so many, in fact, that Martell says it has been impossible thus far to identify the mutation in the cell that causes cancer—if indeed, it is only one combination of changes.

Martell was also pretty sure by 1975 that low-level alpha radiation was responsible, at least in part, for cancer in other parts of the body besides the lung; and that it might be a major factor in causing heart disease as well.

About 20 percent of the insoluble alpha particles that land in the deep lungs, in the pulmonary spaces, or in smaller passages of the lung remain there for about two years. "It is well known," Martell wrote in the *American Scientist* in July-August 1976, that some of these particles move into the lymph system, whose fluid moves them into the respiratory lymph nodes and into the lymph nodes of various organs. Other particles rupture into the blood vessels from the lung, and from the nodes, or knobs, in the lymph system, both of which, in turn, carry the insoluble particles to the liver, spleen, and bone marrow—where they can remain for decades.

This means that insoluble particles wind up at all the sites of the body where human cancers occur—and cancer of the stomach, the esophagus, and other internal organs seem to occur right next to lymph tissue "with visible accumulations of insoluble particles and measurable radioactivity," according to Martell.

"For all these reasons," Martell said in the *American Scientist,* "it seems logical to conclude that internal alpha emitters, rather than cosmic rays or other natural sources of radiation, may be the principal agent of radiation-induced cancer in man."

The implications of this are awesome. For one thing, it would mean that the radiation guides laid down by the government to protect the public health are relatively meaningless in many situations—especially when it comes to the dangers of internal alpha radiation—or when they try to equate alpha radiation with gamma and beta rays. Obviously, the effect of a "whole-body dose" of external gamma rays would be quite different, in consequence, from the same amount of radiation released by an insoluble alpha particle embedded in tissue.

This is just the beginning. There seems to be a good possibility that internal low-level alpha radiation could be one of the leading contributors to atherosclerosis—the chronic disease known as hardening (or thickening) of the arteries, which stymies the flow of blood. And atherosclerosis plays a role in some strokes and every heart attack, in terms of the small growths called plaques that form in the arteries to block the flow of blood. Alpha radiation might be responsible for the formation of these plaques. Martell cites evidence presented by Earl P. Benditt and John M. Benditt of the department of pathology at the University of Washington's School of Medicine in Seattle to indicate that these plaques may be nonmalignant tumors of the artery walls. Attempts to produce plaques by means of chemicals and viruses have been unsuccessful.

But they have been created by irradiating the arteries with X rays and radium. High concentrations of alpha radioactivity have been found in the calcified plaques of atherosclerosis victims.

Smokers suffer heart attacks much earlier in life than nonsmokers. Martell was almost ready to conclude in 1975 that insoluble alpha-emitting particles inhaled from cigarettes could be producing the plaques responsible for these coronaries. It's also worth noting, Martell said, that cigarette smokers not only have a higher rate of lung cancer than nonsmokers—but their rate of other cancers is much higher too. And "the presence of insoluble alpha-emitting particles at these secondary cancer sites strongly reinforces the case for alpha radiation as the carcinogenic agent . . . at these auxiliary sites," according to Martell.

Of course, we're not all cigarette smokers. But all of us come in contact with insoluble alpha emitters in the general environment, which come from many sources —from industry in the form of uranium oxide and thorium oxide, burning coal, smelting lead, from forest fires and burning leaves and crops. We get them in smoke-filled rooms, and from the nuclear fission process, uranium and plutonium in fallout from atom-bomb tests, nuclear reactor plants, transport and processing of nuclear fuels and waste material, making plutonium fuel pellets, and making plutonium triggers for nuclear warheads.

Another fact worthy of notice is that people who live in cities, where there's a lot of air pollution, run a higher risk of cancer, along with people who work where there's a lot of dust—like uranium miners. Martell thinks insoluble dust particles that build up in the lung play an important cofactor role in the incidence of cancer; they clog the flow of lung fluid, which makes it more difficult for the organ to get rid of insoluble radioactive particles. For example, asbestos workers who smoke cigarettes have a bronchial cancer rate that's eight times as high as that of other smokers. But asbestos workers who do not smoke are not more prone to lung cancer than other nonsmokers, according to Martell, who notes that asbestos fibers are highly insoluble, but that they are not radioactive, and do not cause cancer, according to Martell.

Martell also feels that chemical pollutants "may damage lung tissue," which makes it more difficult for the organ to cleanse itself "of insoluble particles, thereby enhancing tumor risks, particularly among smokers. The relative significance of chemical agents, viruses, and radiation in the incidence of lung cancer is not known," Martell says. "And the proposed chemical carcinogens in cigarette smoke and in polluted urban environments have not been demonstrated to be carcinogenic at the low concentration involved." Therefore, Martell concludes it is "likely that radiation, and alpha radiation in particular, may be the principal agent of human cancer."

City traffic seems to play an important role in urban air pollution. Studies released in 1975 indicated that the biggest source of air pollution in some cities such as Chicago is the dust that cars and trucks keep kicking up from the streets. Some of this dust is insoluble. And some of the insoluble dust particles contain alpha emitters, including plutonium 239, which can remain in the environment for centuries and centuries—blowing in the wind, suspended, dropped, picked up and kicked up again by anything that moves—cars, trucks, shoes, bicycles, wagons, dogs, cats . . . and always the wind. Bob Dylan, with the intuition of a poet, set the tone for a generation in 1962 with his anthem, "Blowing in the Wind," which certainly has prophetic implications for today and tomorrow for nuclear fission power and plutonium 239.

In spite of all the allegations about plutonium, many people who believe in nuclear fission power think the dangers of low-level radiation in general and plutonium in particular have been grossly exaggerated. Some have attempted to demonstrate this point by claiming that the toxicity of plutonium is relatively slight compared to botulin toxin, an agent of biological warfare—which prompted Donald Geesaman to retort that no one has suggested burning botulin toxin for fuel.

When I sat down with Congressman Mike McCormack in May 1975, he said that "plutonium is less harmful than cigarette smoke." One member of an environmental group published a letter she received from McCormack in which he added, "You could live your entire life beside a nuclear-power plant using plutonium and be safer than taking a bath." And McCormack's policy papers insist that "not a single person has been injured or killed in any radiation related accident at any nuclear-power plant (or supporting activity)." McCormack also says "the health hazards of plutonium are being fantastically exaggerated by uninformed individuals, and even some scientists.

"We put about 5 tons of plutonium in the atmosphere from nuclear weapons tests. It vaporized. About 4 tons of it came down. The other ton is still drifting around. Everyone in the northern hemisphere is carrying some of this plutonium in his body," according to McCormack. "This is a thousand times more than we thought we could carry without harm. There's no statistical evidence to indicate that one death out of the thousands of people who died of lung cancer last year [1974] was caused by this 5 tons of plutonium. And it's been around long enough to be killing us all if the exaggerated statements of the hazards of plutonium made by the antinuclear activists were even remotely close to the truth."

McCormack's statements from the Capitol infuriated Dr. John Gofman out in San Francisco. "I would love to debate Mike McCormack before a group of scientists," Gofman told me. "What does he expect—that every lung tumor will carry around a little flag saying, 'Look, I got mine from plutonium!'

"The trouble with some people in the nuclear business," Gofman complained, "is that they think their ignorance is knowledge."

Dr. Gofman had time on his hands in 1975. He had

just retired from his teaching job of many years as a professor of medical physics at Berkeley, and had just completed some work he had been doing for a private medical systems firm, so he decided to take another look at the hazards of low-level radiation—which had put him and Arthur Tamplin at such great odds with the nuclear establishment in 1969 when they were still working for the AEC. But this time Gofman decided he would devote all his attention to the hazards of plutonium. He began with the widely accepted assumption that more than one out of every 1000 cases of lung cancer is caused by "natural" or background radiation in America. Then he restudied the work of Donald Geesaman, who in the late 1960s had calculated that there were 90 trillion particles in a pound of plutonium; that one-millionth of a gram in the 454 grams to a pound could cause cancer; and that one case of cancer would occur for every 10,000 particles of insoluble plutonium—which rounded out to 9 billion lung cancer doses in every pound of plutonium 239.

Next, like Dr. Edward Martell, Gofman went back to the U.S. Surgeon General's report of 1964 based on studies of the relationship between cigarette smoking and lung cancer. From this Gofman discovered that a major assumption used to estimate the amount of insoluble plutonium that could be inhaled into the lungs without undue risk was wrong.

The International Council on Radiation Protection's Task Group on Lung Dynamics, and other groups which established the radiation guides adopted in the United States and elsewhere, believed that most of the plutonium particles that enter the lungs are quickly cleared away from the region where most cases of lung cancer occur—at the top of the lungs, in the tracheobronchial (windpipe) region. These radiation authorities believed that only 8 percent of all the plutonium inhaled into the lung was deposited in the upper bronchial region, while 25 percent wound up at the bottom of the lungs in the deep respiratory or pulmonary tissue, where much fewer cases, comparatively, of lung cancer occur. This assumption was based on a faith in the cilia—very fine, hairlike projections in the bronchial system. In constant motion, like delicate fans, they have the important function of sweeping dust and dirt upward and out of the lungs.

The radiation experts said these cilia propelled 99 percent of the plutonium out of the upper part of the lungs in less than a day, sending the insoluble particles toward the intestines and out of the body. They also believed that 40 percent of the insoluble plutonium particles deposited at the bottom of the lungs are cleared away in a day, and that another 40 percent slowly moved back up toward the bronchial tubes, where the cilia pushed them up and out toward the intestine. The remaining 20 percent of insoluble plutonium particles in the lower lungs would be discharged by the lymph system and blood.

But Gofman learned from the Surgeon General's report that cigarette smoking—especially heavy smoking—destroys cilia in the bronchus and impairs the ability of the remaining tiny fanlike projections to function. "All the studies," Gofman said, "including Geesaman's, were based on the assumption that insoluble deposits in the lungs are rapidly cleared by cilia, especially from the tracheobronchial region. But the rate of lung cancer is ten times as high for cigarette smokers as it is for nonsmokers. And if smoking destroys or damages the cilia, then all the assumptions concerning the average rate of insoluble plutonium clearance from the lungs could be vastly in error.

"Everyone was trying to figure out dosages of plutonium in the deep respiratory zone," Gofman continued. "But this is the wrong organ to look at. Only 10 percent of the lung cancer cases occur there. And this led scientists to drastically understate their calculations when it came to estimating the lung cancer hazard from inhaling insoluble plutonium oxide particles."

"What we really needed to have," Gofman said, "was an estimate of the radiation dose delivered by insoluble plutonium particles to the critical tissue in the region where almost all of the bronchogenic cancers occur. And the relevant issue about human lung cancer is the effect of insoluble plutonium particles on 1 gram of bronchial tissue."

Gofman then began his calculations to formulate the risk of lung cancer that we could face from the presence of plutonium in our lives—especially reactor-grade plutonium, which scientists such as Dr. Bernard L. Cohen of the University of Pittsburgh agree is five times as hazardous by weight as pure ("hot") plutonium 239.

These were some of Gofman's conclusions, which the Committee for Nuclear Responsibility published in May 1975: There are over 7 billion potential lung cancer doses in a pound of pure plutonium 239 for smokers, and 42 billion in the same amount of reactor-grade plutonium. For nonsmokers, there are 62.5 million potential doses in a pound of plutonium 239 and 338 million in a pound of reactor-grade plutonium.

Another way of looking at the plutonium hazard, according to Gofman, is to realize that 0.058 micrograms of pure plutonium 239 in the lungs or 0.011 microgram of reactor-grade plutonium represents one cancer dose for smokers, while 7.3 micrograms of pure plutonium or 1.4 micrograms of reactor-grade plutonium is a lung cancer dose for nonsmokers.

Averaging out the risks of lung cancer from insoluble particles of reactor-grade plutonium between smokers and nonsmokers, Gofman concluded that every pound of plutonium that would be used in nuclear power plants would be enough to cause 21 billion cases of lung cancer.

These conclusions led Gofman to charge that the federal radiation guides were much too high, especially for people who worked with plutonium. The "allowable lung burden" of insoluble plutonium 239 for atomic workers who smoke is over four times as high as it should be, according to Gofman, who adds that every smoker and one out of every 30 nonsmokers who worked with or around plutonium "would develop fatal lung cancer at the current permissible dose."

In July 1975 Dr. Gofman published another controversial paper for the Committee for Nuclear Responsibility called "Production of Human Lung Cancers from Worldwide Fallout." At the end of this report, after he finished his calculations on the cancer risks from fallout, Gofman estimated that a fully developed nuclear fission power program would increase the number of lung cancer deaths in the United States by 500,000 a year after the turn of the century. Gofman based his findings on an AEC projection that an expanding nuclear-power industry means we could put over 400 million pounds of reactor-grade plutonium through the nuclear fuel cycle in the United States by the year 2020.

The proponents of nuclear-power plants claim they contain more than 99.99 percent of the radioactive poisons they produce. Gofman doesn't think they do this well, and calculates that even if they manage to contain 99.99 percent of the plutonium produced, the nuclear establishment would routinely be releasing 40,000 out of 400 million pounds into the biosphere, which would in turn be "responsible for 500,000 additional fatal lung cancers a year.

"This would mean," Gofman explained, "increasing the total death rate in the United States by 25 percent each year, since 2 million persons currently die from all causes combined" each year. When he examined the risks from plutonium fallout, Gofman considered the amount of plutonium that had been dispersed into the atmosphere by atom-bomb tests, its measured distribution throughout the northern hemisphere—including the United States—the ratio of smokers to nonsmokers, the fact that it takes 15 years or more for the signs of most cases of lung cancer to appear, and the age brackets—and sex—of people who were exposed to this radioactivity during the years which produced the most fallout.

Gofman cited established percentage statistics reporting that only 20 percent of the 85,000 people who died of lung cancer in this country in 1975 were under 40 years of age and 2 percent were under 50, compared to over 17 percent between the ages of 65 and 70, and about 59 percent between 70 and 80.

Gofman concluded that 116,000 people in the United States will die of lung cancer from plutonium fallout—and 1 million in the whole northern hemisphere. Of this total, according to Gofman, the number of lung cancer deaths "occurring right now in 1975 is probably around 1000 per year in the United States, since the latent period is just about over"; and, that "this number will rise steadily in annual rate . . . over the next couple of decades." Gofman also estimated that some 10,000 persons around the world died of plutonium-induced lung cancer from fallout in 1975. In his report on fallout, Gofman also attacked the pronuclear argument that plutonium has not harmed workers in industry who have been contaminated by it.

There have been studies conducted on several of these groups. One was made of 26 men who worked on atom bombs at Los Alamos, New Mexico, for the Manhattan Project; another was made of 25 workers contaminated by a fire at Rocky Flats. There's not much information about the amount of plutonium that was inhaled by the workers at Los Alamos, according to Gofman, who adds that most of it was probably in soluble form. "Therefore," Gofman said, "at best it would be foolish for anyone to base serious conclusions on the experience of these men." Nevertheless, he took a statistical stab at it, working his percentages on smokers versus nonsmokers and age, concluding that the odds were against seeing any signs of lung cancer among these men by 1975 because it was too early. But there was enough data about the plutonium inhaled by the workers at the Rocky Flats weapons plant, according to Gofman. And he feels that the time to watch for signs of cancer among these workers "will be in the next five to ten years . . . their lung cancer death rate some ten years after 1975 will be of great importance" in attempting to assess the hazards of low-level radiation and plutonium.

Edward Martell has a different slant. For example,

Martell notes that while the medical histories of the Los Alamos workers are usually accompanied by statements that none of their ailments can definitely be blamed on plutonium, he believes the "limited published information . . . is more disturbing than reassuring."

"With equal justification," Martell said, "one may state that most of the serious medical findings in this group can be attributed to plutonium." Only 12 of the 26 exposed to plutonium could have inhaled it, according to Martell, who then presented this look at the group: one of the 26 died in the early 1950s, although Martell does not know why. Another died at age 38 of a coronary. A third recovered from a stroke. A fourth had part of his lower right lung removed by surgery in 1971. A fifth had a malignant tumor in the wall of his chest. A sixth had surgery to remove a bleeding ulcer. One has gout. Another lost his teeth, apparently because the thin fibrous membrane in his jaw was damaged, according to Martell, who said this was the first tissue to show signs of cancer in experiments on beagles that received toxic doses of plutonium. "The full medical history of this group, now mostly in their fifties, has not yet completely unfolded," Martell said, adding that "the medical experience of this small group thus far proves no basis for complacency about the health consequences of plutonium exposure."

Martell's ideas about plutonium fallout risks are also different. His studies have led him to suspect that the most radioactive particles of fallout plutonium in the tissues of the lungs may have the right dosage to kill all the cells within the alpha range; but that its rate of solution is too low to produce tumors in cells at a greater distance. What this means, according to Martell, is that the radiation dose from the hottest particles of fallout plutonium provides radiation therapy; that is, it's more likely to wipe out cells—including mutated cells within its range—than to trigger a tumor. He suggests that uranium and the plutonium in fallout of lower radioactivity carry a higher risk because they can cause irreversible structural changes producing mutated cells that will live to proliferate into a tumor.

Martell used to think that hot plutonium 239 with its high concentrations of radioactivity also had the same "therapeutic" effect, which would mean that a hot particle of plutonium would simply "overkill" all the nearby cells. But pure plutonium oxide is much "hotter" than fallout plutonium; its radiation is much, much stronger—and a billion times as intense to surrounding cells as natural dissolved alpha radiation.

Martell claims that the radiation from an insoluble particle of pure plutonium in the lungs does two things: it kills the cells within the normal alpha range; but then its dissolved activity spreads outward in all directions beyond this normal alpha range to irradiate the rest of the cells in the lung with elevated doses of alpha radiation which cause the structural changes that can lead to cancer. In Martell's words, "This secondary dissolved activity gives rise to the tumors from 'hot' particles of plutonium."

The implications of Martell's interpretation are extremely disturbing. There are about 50 million cells in a gram of lung tissue within the alpha range of insoluble particle burdens. These are the cells that are attacked by insoluble alpha particles such as polonium 210 in smokers. But there are 50 billion cells in the whole lung. Which means that while only one thousandth of the lung is exposed to radiation from alpha particles that cluster together, way over 49 billion cells are irradiated by dissolved activity from hot plutonium embedded in tissue.

Therefore, if Martell's views are correct, this dissolved alpha activity from hot plutonium, which delivers levels of alpha radiation needed to produce the mutated cells that could lead to cancer, is attacking billions of cells in the lung, rather than millions. This is why Martell says the plutonium cancer risk for the general public and for those who work with or around plutonium is "unacceptable."

And we know that plutonium has escaped to the environment, and that workers have been "contaminated" by plutonium. Let's take this report from Rocky

Flats, for example, which called attention to the dangers of low-level radiation.

In 1970 testimony was presented before a Colorado Industrial Commission hearing, when Dow Chemical was still operating Rocky Flats, to indicate that some workers had 20 times as much plutonium in their lungs as the federal radiation guidelines allowed, with Dow Chemical spokesmen claiming this was still 0.1 percent or less of the level of harm. At the same hearing, a Dow medical director said 56 out of 4700 persons employed at Rocky Flats between 1953 and 1970 were known to have had cancer—that's about one out of 80 people, compared to about one out of 600 for Colorado in 1970 who were roughly in the same age range. And scientists like Martell say there may have been many, many more cancer victims at Rocky Flats—and some who died of heart attacks before they could get cancer.

Of the 56 who had cancer, 48 percent worked in plutonium processing buildings, according to Dow as reported in the *Denver Post,* which covered the hearings in August 1970. Fourteen of the 56 died, of whom 6 had worked near plutonium. The Dow medical director said 2 were killed by cancer of the brain, and that 1 each died of cancer of the lung, colon, urinary tract, and bone.

These figures might be just an indicator, pointing at the tip of the relationship between plutonium and cancer if the findings of Edward Martell and John Gofman hold up. But the people I talked to at Rocky Flats in 1975 dismissed allegations that the plant might be a hazardous one to work in or live near. Alan K. Brown, a chemist at Rocky Flats, told me about the safety systems and devices to control plutonium and said the amount of radioactivity inside the plant and released to the outside was only a fraction of the radiation levels established by the Environmental Protection Agency. And James R. Nicks, one of 60 persons employed by the federal government at the plant to oversee its operations, said, "To a man, this is not a hazardous place to work. There's zero turnover. The only reason people leave here is because of retirement. We have logged millions and millions of safe man-hours at Rocky Flats. The only person who got hurt last year was someone who slipped on the ice in a parking lot."

In response, Martell, looking toward the Rocky Flats plutonium plant from the windows of the National Center for Atmospheric Research, said, "People who work there have faith in institutions—in what the federal government tells them. And the government tells them it's safe to work there. So why should they give up their good paying jobs? Besides, cancer is slow in developing. In 15 years or more, after it develops, no one will know for sure exactly what caused it."

The dangers of low-level radiation were publicized at another plant where workers were contaminated by plutonium. This was the Kerr-McGee Corporation's nuclear fuel manufacturing facility along the Cimarron River near Crescent, about 30 miles north of Oklahoma City—which attracted a great deal of notoriety stemming from the mysterious death of Karen Gay Silkwood on November 13, 1974, who was on her way from Crescent to meet a reporter and a labor union official in Oklahoma City when her car ran off a straight road and crashed into a culvert. State and federal investigators said her death was an accident. But some people think she was murdered—that her car was bumped from behind and forced off the road. And a packet of documents Karen Silkwood reportedly compiled to take with her to support her charges about alleged plutonium dangers at the plant—which the two men waiting for her were expecting to receive—has never been found.

Prior to her death, the 28-year-old laboratory technician at the plant kept saying workers were being endangered by radiation because Kerr-McGee's health and safety procedures governing plutonium were either ignored or too lax; she also charged that inspection reports were falsified on plutonium fuel rods intended for use someday in an experimental fast-breeder test reactor. Ms. Silkwood was one of the workers at the plant who was contaminated by plutonium—once in the late summer of 1974 and again on November 5, 6, and 7. Traces of plutonium were found in the apartment she shared with a 21-year-old coworker named Sherri Ellis who was also contaminated. Although there have been a number of morbid guesses, no one has been able to say for sure how Karen Silkwood and her roommate were contaminated, or how a thousandth of an ounce of plutonium got out of the plant and into her apartment—a factor which also cast doubt on Kerr-McGee's plutonium safeguards.

The last five months of Karen Silkwood's life seem to have been exceptionally troubled. She was divorced and her three children were living with their father. The reverberations of the counterculture spawned in the 1960s—the music, drugs (hard and soft), and the life-style—apparently had an unsettling effect on Ms. Silkwood. She also became a militant member of the organized labor union that represented workers at the plant and was caught up in the bitter clash between Kerr-McGee and the Oil, Chemical, and Atomic Workers International Union.

Karen Silkwood had been an honor student in school and apparently had a reputation for being conscientious and willing to fight for what she thought was right. Although she had been pro-union from the time she went to work at Kerr-McGee in 1972, there is little to indicate that she was overly concerned about the dangers of plutonium until the summer of 1974—even though 77 employees at Kerr-McGee had been contaminated in 17 incidents since the plant began to operate in 1970. Also, there had been reports that radioactive waste had been shipped improperly, and drums containing waste had leaked.

In the late spring of 1974 Karen Silkwood was elected to the three-member governing board of the union local at the plant: her concern about the health and safety of her coworkers at Kerr-McGee who handled plutonium then began to quicken.

Mrs. Ilene Younghein of Oklahoma City, one of the antinuclear leaders in her state, had this to say:

"A turning point in the whole operation occurred when Dr. Dean Abrahamson and Don Geesaman came down from the University of Minnesota to speak at a union meeting in late summer of 1974. Both men, experts on plutonium, advised the employees of the serious implications of overexposure. Evidently the company's training sessions hadn't spelled out the hazards very well. It was at this time that Karen Silkwood, along with other employees, became very concerned about violations of safety procedures at the factory."

On September 26, 1974, Karen and two coworkers flew to Washington, D.C., to tell union officials how they believed operations at Kerr-McGee were endangering the lives of workers at the plutonium plant. Steven Wodka, a union worker, took the Oklahoma delegation over to AEC headquarters at Bethesda, Maryland, where they also presented their story to the agency. Around the same time, the three workers from Kerr-McGee told the union that they believed the plant was manufacturing faulty fuel rods and that inspection records had been falsified.

Between her meeting in Bethesda in late September, to discuss the alleged dangers and irregularities at the Kerr-McGee plant, and her death in November, people close to Karen Silkwood said she was obsessed by the dangers of plutonium and union-management troubles at the plant. She began to take copious notes about the plutonium factory. She was also taking physician-prescribed pills to sleep her tensions away.

Then she was contaminated by plutonium three days running in early November. Karen Silkwood underwent continual medical examination and procedures to cleanse her of plutonium. On November 7, the third day, health-physics personnel went to the apartment she shared with Sherri Ellis and discovered it was radioactive from plutonium. By this time Karen Silkwood was apparently close to hysteria: she knew enough by now about the dangers of low-level radiation from plutonium to realize how hazardous it was. Karen Silkwood spent the last days of her life talking to investigators from the AEC and the Oklahoma State Health Department, who were trying to find out how she had been contaminated—and to what degree, and how plutonium had gotten out of the plant and into her apartment.

On November 13, 1974, the day she died, Karen Silkwood met with members of her union local in Crescent. An affidavit taken after her death swore she was clutching a notebook and a manila folder containing what Karen Silkwood said was "the truth" about Kerr-McGee's plutonium plant. When the meeting was over, Karen Silkwood said she was on her way to turn her materials over to Steve Wodka of the union's Washington office and a reporter from the *New York Times* Washington bureau named David Burnham, who had been writing about nuclear power.

Wodka and Burnham were waiting for her at a motel on the outskirts of Oklahoma City. Karen Silkwood drove off from Crescent in her 1973 Honda to meet them. But her car went off a straight highway around 7:15 p.m. and crashed into a culvert 7 miles south of Crescent. Karen Silkwood was killed instantly. State police said she had fallen asleep behind the wheel. An autopsy report said there was a sedative named methaqualone in her blood when she died.

The Oil, Chemical, and Atomic Workers International Union hired an accident expert from Dallas, A. O. Pipkin, Jr., to investigate the crash. He reported back to Anthony Mazzocchi, a legislative director for the union in Washington, who on the strength of Pipkin's report sent a telegram to the Justice Department calling for a thorough investigation into the Silkwood death. Mazzocchi said the union had "evidence to suggest that Ms. Silkwood's car was hit from behind by another vehicle, causing her car to leave the road and hit the concrete culvert."

The FBI launched a probe, then said it couldn't come up with anything to support the allegation that Karen Silkwood had been murdered. But the union kept pressing for a thorough, independent investigation. Feminist and citizens' action groups accused the government of an "official cover-up." They wanted to reopen the case, which has enough unexplained quirks to whet any conspiratorial appetite. And by 1975, responding to pressure from citizen groups, the Senate Government Operations Committee said it would conduct a congressional investigation into Karen Silkwood's death.

Kerr-McGee's immediate troubles did not end with the death of Karen Silkwood. On December 17, 1974, for example, it was reported that five more workers at the plant had been contaminated by radioactivity. And the next day, the company said it found some uranium fuel pellets outside the plant. On December 29, 1974, *New York Times* reporter David Burnham published a story saying that Kerr-McGee had been unable on occasion to account for plutonium at the Cimarron plant. There were times, according to a company executive, when 60 pounds of plutonium were missing during periods when the plant contained up to 600 pounds of this radioactive material. And federal files revealed that Kerr-McGee had been reprimanded a number of times for alleged failures to keep track of plutonium. One report said the company had to close the plant in 1972 in an effort to track down 50 missing pounds of plutonium—much of which seems to get lost in machinery. The missing plutonium, of course, called more attention to the dangers of low-level radiation and to the issue of health and safety at the plant.

In early January 1975, AEC investigators substantiated union allegations that X rays of fuel rods and other records had been falsified; and the agency found that 20 of 39 union charges about health and safety violations at the plant had merit either in whole or in part. Finally, in October 1975, 11 months after Karen Silkwood's death, Kerr-McGee announced it was going to shut down its plutonium fuel fabrication plant near Crescent, saying its decision was based on a failure to obtain new contracts.

When I talked by phone to several reporters in Oklahoma City who covered the Karen Silkwood story, they contended that her death was blown out of proportion by the national press; that she was a trifle "kooky"; that she was not murdered; and that there was not much to worry about at the Kerr-McGee facility—although these reporters did accuse the company of being less than candid about operations at the Cimarron plants. Mrs. Ilene Younghein of Oklahoma City, who is very critical of nuclear power in general and Kerr-McGee in particular, says, "I really doubt that Karen Silkwood was murdered," hastening to add that this didn't mean she was defending the company, whose influence she called "extremely powerful" in Oklahoma. Certainly the three newspapers in Oklahoma City seem to have cordial relations with Kerr-McGee. In 1975, for example, the *Daily Oklahoman* called Karen Silkwood "an antimanagement agitator whose accusations against Kerr-McGee have been largely disproved." In another editorial the paper said, "The only safe, clean, nonpolluting, and plentiful alternative fuel is plutonium . . . That is why banning plutonium from Oklahoma would be like the ostrich sticking its head in the sand at the approach of danger."

The *Daily Oklahoman* had earlier claimed that "this state stands in the forefront of the nuclear-power industry today because of the hundreds of skilled workers who are already at work processing nuclear fuels into usable forms, and the sophisticated plants in which they are employed."

In spite of all the publicity about the hazards of low-level radiation, people continue to work with and around plutonium at places like Rocky Flats and the Kerr-McGee plant. Anthony Mazzocchi, the legislative director for the Oil, Chemical, and Atomic Workers International Union—which represents about 20,000 atomic workers out of a total union membership of 200,000, said dangerous conditions wouldn't deter people who are desperate for work. "Our people are caught in the real situation of needing jobs," he said. "They have to eat. And they figure that there's always a chance they won't get cancer—that it will get the next person, but not them. The trouble with this society is that it's totally profit-oriented," Mazzocchi added. "Everything is produced for profit. If we had a rational society, people wouldn't be driven to seek out dangerous jobs. Perhaps the fundamental point is that maybe this society doesn't really work."

The labor union official's words stayed with me as I kept asking Dr. Edward Martell, during our day-long conversation in Boulder, and later, during our many talks on the phone, and a second meeting in Chicago, why, from what I could gather, relatively few scientists seemed to be aware of the implications of his work on internal low-level alpha radiation, cancer, and heart disease. He kept repeating what he told me the first time we met—that the National Cancer Institute with its multimillion-dollar budget was largely ignoring the possibility that low-level radiation might be responsible for most human cancers because the institute assumed that the AEC—now the Nuclear Regulatory Commission, and the Energy Research and Development Agency—was taking care of all necessary research connected with the dangers— known and potential—of radiation.

"The fact that the National Cancer Institute does not have a program of radiation-induced cancer research is a surprising and disturbing state of affairs considering the fact that nuclear radiation is known to be a very effective agent of cell mutation and cancer; and that the number of cancers due to radiation compared to that of other cancers is not known," Martell said.

In defense of the NCI, Dr. Gio Batta Gori, a bacteriologist and physiologist who is the deputy director of the Institute's Division of Cancer Cause and Prevention, in March 1976 told me "it would be foolish for us to do research on radiation and cancer; it is not necessary, because everyone already knows that radiation causes cancer. The important question," according to Dr. Gori, "is how much radiation is in the environment, and what effect it has. But we're not equipped to police the environment for radiation; nor have we been given the mandate to do so."

When it comes to Martell, Gori agreed that Martell has a "viable theory," one which Gori said the NCI, acting upon the recommendation of a panel of consultants, is now in the process of investigating—although "not [Gori admits] with exceeding great enthusiasm" or at great expense; and what the NCI is studying is Martell's earlier work, limited to natural alpha radiation, cigarettes, and lung cancer.

"What about the nuclear scientists who work for the government, or do contract work for the government. Don't they know what you're talking about?" I kept asking Martell.

"Sure," he said. "Some of them do. But they don't want to endorse, or even acknowledge, the credibility of a hypothesis that's going to put them out of business—at least when it comes to nuclear fission power."

Martell told me that scientists employed by the NRC and ERDA have gone over the material he's published.

"Has anything happened as a result?" I asked.

"Nothing," Martell replied. "They don't say a word. They just want to ignore my articles in *Nature* and the *American Scientist*. Some of these scientists seem terribly disturbed. They know what I'm talking about. But they haven't made a single move to refute me, to say how or why my hypothesis about the threat of low-level radiation from insoluble alpha particles could be wrong; to tell me why it's not responsible for lung cancer in smokers and probably other cancers and heart disease as well. They want to ignore such possibilities, and hope that the subject will just go away.

"The nuclear proponents have been trying to shove nuclear power down our throats—safe or not—for years," Martell added by word of explanation. "They're not about to stop now."

I spoke with one senior member of ERDA's staff who flew to Boulder to spend the day with Martell. He is Dr. Nathaniel F. Barr, a chemist, who is one of the top men in ERDA's division of environment and safety. "It's hard not to be fascinated by Martell," I was told by Barr. "He's enthusiastic and a very capable man. And Martell's working in a very interesting area, approaching it in a very interesting way. He's deliberate and responsible and his radiochemistry is very fine. But all Martell has is a postulate—a theoretical possibility, but no evidence to support his case. No evidence that alpha radiation causes cancer."

And if Martell is right? I asked.

Here we danced around a bit.

"Then the standards on plutonium must be wrong," Barr said in response to a direct question. Even so, Dr. Barr insisted that plutonium would not represent a significant risk for the public health and safety because it would "increase this [our] exposure [to alpha radiation] by a very tiny fraction of what now exists." Barr added that he was sure Martell, in time, would get the comments he was looking for "because he's presented his work to the scientific community in the right way."

Dr. William J. Bair, a radiobiologist, is another one of the scientists I approached about Martell's work. Bair is the manager of Battelle Pacific Northwest Laboratories' biomedical and environmental research program in Richland, Washington, which is doing plutonium research work under an ERDA contract.

Bair said Martell has an "interesting hypothesis, but I find it hard to believe. I don't believe that he has the evidence to support it. I just don't happen to agree with him."

Again, what if Martell's right?

"I don't think it would rule out the use of plutonium," Bair said. "We have the technology around to control it—to incorporate it. It would not represent a real increase in the amount of plutonium around." Bair stressed, at this point, that he had a "high respect for the toxicity of plutonium—I've worked with it for 20 years, and wouldn't want to inhale any." Bair, who feels we "have to have nuclear power," was kind enough to supply me with a generous amount of reading material dealing with plutonium, its effects and potential risks. I read it all with care and appreciation.

Certainly I am not a scientist, nor can I pretend to anything approaching expertise in this complicated field, but I couldn't find much in this reading matter that seemed to weaken Martell's case; I did find a lot that seemed to strengthen it—including information about the residence times of plutonium in the body, the body's clearance mechanisms, the behavior of soluble versus insoluble particles of plutonium, the activity of uniform and nonuniform distribution of plutonium, and the wasted radiation—or "overkill"—of the local cells from "hot" particles of plutonium.

Another scientist doing plutonium research work under an ERDA contract is Dr. Otto G. Raabe, a biophysicist with the Lovelace Foundation's Inhalation Toxicology Research Institute in Albuquerque, New Mexico. "Martell's hypothesis is not totally accepted," Raabe said. "It's based on limited experimental work. The activity of radiation involved is very low. I'm skeptical. His hypothesis is not very good. But it cannot be disproved. It could be a direction for future work . . . He's a very smart fellow, a good scientist, but there's just not enough data—to prove or disprove his case. . . . If Martell's right, then the standards [governing plutonium] are wrong. If he's right, then we have a real serious problem. But we're going to have to move ahead with nuclear power," Raabe said. "It has to be the most important source of power in the next 100 years. It has the best record of safety of any industry. Certainly there's room for caution," Raabe added.

And Martell is very cautious. His work in the laboratory has always been ahead of the findings he has published. In the fall of 1975, he told me this: "I expect that in about a year and a half from now, they [scientists] won't be able to ignore the stronger evidence we should have by then. And that should be the end for nuclear fission power. If low-level alpha radiation is already mainly responsible for three-fourths of all nonaccidental deaths in the United States, you just can't permit the use of an alpha emitter like plutonium. What it will mean is that nuclear-power reactors are totally unacceptable." And, Martell added, that nuclear war is "unthinkable."

In 1976, Martell added this footnote: "I have a

well-justified hypothesis. And the pieces keep falling into place [to support it]. I'm not surprised by the reaction, from the built-in resistance of people who can understand but don't want to. But I'm confident. And step by step I'm going to the scientific and medical community with the evidence. The serious risks I'm attributing to internal alpha emitters can be evaluated experimentally," Martell stressed. "And, in my opinion, it's essential that we do so before we commit ourselves to nuclear power."

Perhaps people like labor union official Anthony Mazzocchi and Martell are right—that there are too many people in our society who have been blinded by self-interest or the profit motive to the extent that they've been willing to ignore serious threats to the public health and safety. But even before Martell and Gofman came up with their latest findings there was surely this to trouble our native tradition of common sense, of which we speak with such pride: There are 454 grams in a pound. Inhaling 0.001 gram of plutonium into the lungs will launch an immediate deadly assault on the tissues. One millionth of a gram, and perhaps far less, can cause lung cancer. One large contemporary light-water reactor produces between 400 and 600 pounds of plutonium 239 a year. A large-scale sodium-cooled fast-breeder reactor would contain a couple tons or more of plutonium.

If the nuclear establishment has its way today, there will be millions of tons of plutonium moving around the United States tomorrow, in fuel fabrication plants, reactor power plants, reprocessing plants, waste storage facilities, trucks, trains, and planes. What we're talking about, then, is pounds . . . tons . . . and then, a lethal millionth, even a trillionth of a gram of plutonium.

Some scientists and critics warn that we are in the dawn of the Carcinogenic Century—the Age of Cancer. Certainly we know that cancer is on the rise. Especially lung cancer, according to the American Cancer Society. John Gofman says we have reached the point in time when plutonium spawned by the atomic age and released into the environment during the atom-bomb testing days of the 1950s and early 1960s will begin to claim its victims. And Edward Martell's findings about the dangers of low-level radiation could be the clue about the cause of cancer —and heart disease—that scientists have been looking for.

VI
The "Hot" Fuel Cycle

I was in Hanford, Washington, trying to gain an over-all perspective on the trouble-plagued nuclear fuel cycle in the United States. Lyle Wilhelmi, my nuclear host and guide for the day, put a dime into the slot of what looked like a gumball machine, and turned the handle. A small white envelope bearing the label "Atomic Marble" came sliding out.

"Here," he said, handing it to me. "It's perfectly harmless."

I read the writing on the packet, which said, in part: "From the U.S. Energy Research and Development Administration's Hanford Plant, Richland, Washington . . . This glass marble was colored by exposure to gamma radiation from a cobalt 60 source used for developing the peaceful atom. . . . The marble is not radioactive and is perfectly safe to handle. . . . Souvenir of the Hanford Science Center."

The light brown marble, like everything else in the nuclear museum of which Wilhelmi was the director, was meant to be reassuring. There were even a few oil paintings, serene and fairly large, of nuclear-power-plant reactors hanging on the walls of the center when I was there in June 1975. "A local artist did them," Wilhelmi volunteered. "She moved here about a year ago, and wanted to start doing pictures that would have some local interest."

No wonder. It would be difficult to find a more pro-nuclear community in the United States than this 90,000 population tricity area of Richland-Pasco-Kennewick in southeast Washington that borders the Hanford nuclear complex, which, in 1975, represented a total capital investment of close to $2 billion, employing around 8500 workers on an annual payroll of some $300 million. It is a community that seemed to depend on funding from the federal government; and the people here were represented on Capitol Hill by hard-driving Congressman Mike McCormack, who came to the House after a long career as a research scientist for the Atomic Energy Commission at Hanford.

Wilhelmi and I continued our tour of Richland's atomic museum, where a visitor can learn almost everything about nuclear fission power, except that it can be terribly dangerous, in spite of the fact that the history of the atom bomb was chronicled by large photographs and bold text displayed in the center. The museum has clearly gone to great lengths to persuade the public that there is a vast difference between the "destructive atom" and the "peaceful atom"—as if they were distant cousins, rather than brother and sister, born of the same nuclear fission process. But many critics of nuclear plants and their related facilities in the nuclear fuel chain contend they represent the same kind of threat to humanity, in their own way, as the atom bomb does. Take the case of a fuel reprocessing plant—which would contain radioactive waste drawn from many nuclear plants. One of these large facilities, after five years, according to Dr. John Gofman, could hold "approximately the radioactivity that would be left decaying for hundreds of years from a large full-scale nuclear war." No one know how we're going to dispose of high-level radioactive nuclear waste, of which there are millions of gallons in temporary storage at Hanford. But here, at the Science Center, every effort had been made to present the "peaceful atom" in a benign light: bright displays of fuel pellets and rods; a large, wall-sized outline of the history of the atom in general and Hanford's role in particular—which produced the plutonium bomb that exploded over Nagasaki; large-scale models of nuclear-power plants, fuel reprocessing plants, uranium fuel enrichment facilities, and radioactive waste storage systems. There are also movies, slides, and bundles of literature and imitation souvenir fuel pellets for the public.

Wilhelmi, an enthusiastic man with a wide range of interests, was a biologist who had been a science teacher in the Oregon schools before coming to Richland in 1975. He was very proud of the atomic museum and was full of plans to expand its scope by giving the thousands of people who visited the center each year a good look at other new sources of power such as solar energy.

But for the moment the museum was going all out to tout the development of the new liquid-metal fast-breeder reactor, which the Ford administration was calling the nation's number one priority to solve the energy crisis—if indeed, there really was one in 1975. I paused in front of a sign on the wall at the center, quoting former President Richard M. Nixon, who in his National Energy Policy Statement of June 4, 1971, said, "Our best hope today for meeting the nation's growing demand for economic clean energy lies with the fast-breeder reactor." And a booklet published by Westinghouse Corporation, which Wilhelmi handed me, called the breeder reactor "our only reasonable alternative . . . essential if this nation is to meet energy demands and become self-sufficient in the decades ahead."

Westinghouse, of course, was building a fast-breeder reactor test facility out on the sagebrush desert at Hanford for the federal government, and the giant company also hoped to be among the builders of a large demonstration breeder reactor plant on the Clinch River near Oak Ridge, Tennessee.

The nuclear critics were trying to slow down and stop the development of the breeder. Most opponents of the controversial breeder said it was too dangerous in every respect; others claimed that even a slew of breeders across the country by the turn of the century couldn't begin to supply us with the energy we might need by then, and there were many people outside the circle of critics who agreed with them that developing the breeder could be prohibitively expensive. There was also the special concern that pushing ahead with the breeder might saddle us with unsupportable risks born out of the dire necessity of trying to protect plutonium from theft.

One point, however, that everyone, pro and con, seems to agree on, is the fact that nuclear fission power could grind to a dead end by the turn of the century (if not before) without the use of the breeder reactor on a large scale. Contemporary light-water reactors run on enriched uranium 235, of which the supply is dwindling; and low-cost, high-grade uranium ore may be all gone, at least in this country, by the year 2000 or soon thereafter, which would tie up and close the contemporary nuclear fuel cycle at its very source. Some utilities were already canceling their orders for reactors in 1975 because they felt uranium was in short supply and too expensive. This is why there is such insistent pressure from the atomic establishment in behalf of the breeder reactor, which can be programmed to produce electrical power and large amounts of plutonium to take the place of uranium as the principal fuel in the reactors of tomorrow. In fact, without the breeder, according to a spokesman for the Federal Energy Administration, there will be "a rapid decline of the nuclear industry."

In 1974, government officials were admitting that the demand for uranium—expected to triple by 1980, and septuple by 1985—was going up so fast that we might not be able to produce enough in the United State to meet our demands by 1985. And the market price of uranium oxide concentrate—called yellowcake—which had fluctuated between $6 and $8 a pound in the early 1970s, bounced up to $15 a pound by the beginning of 1975, then more than doubled to $35 a pound at the end of the year. By 1980, the price of uranium concentrate was expected to reach $47 or more a pound, I was told in 1976 by George White, Jr., vice-president of the Nuclear Exchange Corporation of Menlo Park, California, the principal uranium fuel brokerage house in the world.

At one time, when high-grade domestic uranium seemed plentiful, government spokesmen claimed that nuclear-power reactors would be much more economical to operate than fossil fuel plants—once the capital costs of construction were behind—because uranium fuel would be much cheaper than oil or coal. But they qualified this by saying that "the economics of nuclear power go way down" if uranium ever reached a cost of $30 a pound. This position changed. Even though an Energy Research and Development Administration estimation in 1975 said that uranium had to cost less than $16 a pound to compete with coal—then priced at $20 a ton—some staunch nuclear advocates in the government began to argue that uranium at $60 a pound would still be economically competitive with fossil fuels.

The nuclear critics do not agree. James E. Duree, an attorney from Westport, Washington, a member of the Sierra Club and Friends of the Earth, one of the best-informed, most aggressive critics of nuclear power in his home state, put it this way in 1975: "With the price of the fuel quadrupling in two years, the end of high-grade sources of uranium in sight, and foreign sources either firmly in the control of a cartel, or perhaps inaccessible at any price, the atomic establishment is headed down a street without joy at best, and a dead-end street that terminates at disaster at worst."

In the late forties and early fifties, prospectors who swarmed around the West and Southwest armed with Geiger counters were able to come up with rich deposits on outcropping rock formations. But these are long gone; half of the uranium is now mined deep below the surface of the earth, and much of the remaining uranium in this country is covered by water, glacial drift, and other difficult material barriers.

There are, for example, vast amounts of uranium shale in eastern Tennessee; but this might be prohibitively expensive to mine because it would take 100 times as much of this rock to produce the same tonnage of uranium yellowcake powder as Western sandstone now yields. And it might take more energy to extract this uranium than we would ever get back from it in a reactor. Around 13,000 usable tons of uranium are mined each year in the United States, primarily in Colorado, Utah, and Wyoming. The net has been around 4 pounds of yellowcake per ton of ore. But this ratio is expected to drop, as high-grade ore grows scarce, to about 2½ pounds per ton by 1985, which means that the current capacity to process 28,000 raw tons of ore per day might have to quadruple by 1985 to keep up with demand. By that time, according to uranium brokers, the price of uranium may go up to $50 or $60 per pound.

This could pose a perilous crunch for nuclear-power plants because the reactor core of a 1000-megawatt plant contains about 100 tons of uranium dioxide in fuel assemblies, a fifth to a third of which, depending on the type of light-water reactor, has to be replaced each year.

Westinghouse was already hit hard by the short supply and increasing price of domestic uranium. In the late 1960s, when uranium seemed plentiful and the price was low, and the manufacturer was trying to boost its reactor sales, Westinghouse sweetened its contracts by agreeing to supply its customers with uranium at a fixed price. As 1975 was drawing to a close, Westinghouse was saying it didn't have enough uranium to fulfill some 20 contract commitments that it had signed, and that the company could lose as much as $1.5 billion if it had to buy uranium on the open market to honor its utility obligations. And the utilities, in turn, filed suit to force Westinghouse to fulfill its contracts.

Some utilities were even buying their own mines and mills in an effort to protect themselves. Commonwealth Edison in Chicago, the nation's largest nuclear utility, bought its own uranium mine and mill for $18 million in Colorado. The giant government-owned Tennesseee Valley Authority, which bought a uranium complex in South Dakota, also signed a contract for millions of pounds of raw uranium ore from Canada.

And there were other indications that American nuclear plants would be looking for uranium imports to fuel their reactors. Washington, in turn, is going to lift the current embargo on uranium. In 1977, American utilities will be able to begin loading their reactors with 10 percent imported fuels, a ratio that will continue to escalate upward until it reaches 100 percent by 1984.

A uranium cartel has already been formed to take advantage of the demand for uranium. It is called the Uranium Producers Forum. This cartel is composed of 16 mining companies in Britain, France, Australia, Canada, and South Africa. Most of them are controlled by the French and British Rothschilds. The clan denies allegations that it is a cartel, preferring to call itself a trade association. Whatever the name, the world price of uranium managed to quintuple since the members first met in South Africa in May 1972. The cartel reportedly controls enough reserves of high-grade ore that it can literally move the world price up and down, like a yo-yo on a string, simply by withholding or suddenly flooding the market with uranium, which could squeeze American firms that mine, mill, process, and consume uranium.

This has sparked sarcastic comments from the critics, who say that nuclear power will hardly make the United States energy-independent, as its advocates so staunchly claimed, if we have to depend upon foreign fuel for our uranium-hungry nuclear-power plants. And the uranium 235 that our contemporary nuclear plants use as fuel has to be 3–5 percent enriched. There will be a shortage—and soon—of enrichment facilities to do this job.

At present, there are only three nuclear fuel enrichment plants in the United States—government-owned but operated by private industry. Union Carbide runs the enrichment plant at Oak Ridge, Tennessee; Goodyear Aerospace is responsible for the other two, at Portsmouth, Ohio, and Paducah, Kentucky. These gas diffusion plants represent a total investment of 2.4 billion taxpayer dollars. By 1975, the government and industry were estimating that it would cost about $3.5 billion to build just one gas diffusion enrichment plant —which moves uranium hexafluoride gas through a series of stages, each of which increases the percentage of uranium 235 in the mixture.

The uranium enrichment plants are huge affairs; they sprawl out over hundreds of acres, circulate millions of gallons of cooling water a day, and consume a tremendous amount of electricity—enough, by conservative estimates, to supply a city of a half million people with power.

Some critics, among them David Comey, note that these three enrichment plants consume about 3 percent of all the electricity produced in the United States. The net amount of electricity produced by reactor plants in 1975 was thus 5 percent rather than 8 percent of the country's total. And these enrichment plants are superpolluters, according to nuclear critics like Comey, because they burn low-grade coal for power and discharge vast quantities of carbon dioxide, sulfur dioxide, and soot through their unfiltered stacks.

Most of the enriched uranium produced by these three plants in the past was used for atomic weapons; now the majority of it is purchased for nuclear plants. There are plenty of customers; in fact, with 55 plants operating in this country in mid-1975, 63 more under construction, and dozens of others on order, demand has begun to outstrip the production capacity of our enrichment plants and some utilities are now exceedingly anxious, especially those whose plants will begin operating after 1982, because there is no guarantee that there will be enough enriched uranium fuel for the new reactors.

The three government plants, which stopped accepting new orders in June 1974, are booked solid for the next 25 years. Many of their customers are overseas; in fact, about 80 percent of the reactors in Europe buy their fuel from the United States—an export business which brought in around $421 million in 1974 and could reach an annual figure of $5 billion in the 1980s.

Washington figures we will need at least ten more enrichment plants before the turn of the century to keep up with demand, costing well over an estimated

$30 billion to build. President Ford wants private industry rather than government to build and run them. Consequently, on June 26, 1975, the president asked Congress to end the government's 30-year monopoly on the enrichment process—a matter of great concern and tight security because an atom bomb could be built with 37 pounds of very highly enriched uranium.

Mr. Ford rationalized his request by arguing that the new customers would be civilian rather than military as they had been in the past. But members of Congress said they would have to be assured that private industry would be able to maintain stringent safety and security standards in the manufacture and shipment of fissionable materials before they would agree to go along with the president. Private industry, however, has been reluctant to invest in enrichment plants. Many industry spokesmen say the costs will be extreme for a venture that seems so clouded by uncertainties. At best, they note, it would mean investing in a project with a 30–40-year lifetime. By that time we might run out of uranium to enrich, and the use of plutonium-fueled plants could undercut the demand for uranium before then. But there are firms that have been interested in enrichment plants. Bechtel Corporation of San Francisco, which constructs nuclear plants, is one of them. In 1972, Bechtel put together a consortium called Uranium Enrichment Associates, with Westinghouse and Union Carbide; the latter two firms pulled out of the venture in 1974. Then Goodyear Aerospace, in 1975, said it would join Bechtel and foreign investors—possibly from Iran, West Germany, and Japan—to build a $3.5 billion enrichment plant on a 1720-acre tract along the Chattahoochee River near Dothan, Alabama.

This proposed gaseous diffusion plant, which could supply over 90 nuclear-power plants with enriched uranium fuel, would take about eight years to build; and its federal and private backers hope it could begin operating by 1984. If the proposed project goes through, Bechtel and Goodyear would be protected against loss because President Ford in June 1975 asked

Congress to buy back the plant—and the next two built —at a cost not to exceed $8 billion if the venture fails because of circumstances beyond the control of the proposed builders.

Other combines might enter the fuel enrichment field, too, if Congress votes to end the government's monopoly on uranium enrichment in the United States, agrees to underwrite new facilities, and permits foreign investors to participate in what has been a secret process for the past 30 years. Among the interested parties are Centaur Associates, a consortium of Atlantic Richfield and Electro-Nucleonics, Inc., of Fairfield, New Jersey, Exxon Nuclear Company, of Bellevue, Washington, and the Garrett Corporation. These organizations have expressed interest in the new centrifuge enrichment process, which, in common with the diffusion process, handles uranium as hexofluoride gas to separate and concentrate the lighter atoms of uranium 235 from the dominant mass of uranium 238.

The breeder reactor, so dear to the hopes of the nuclear establishment, will in time reduce or bypass the tremendously expensive necessity of enriching uranium 235. But developing the breeder can turn out to be very costly, too. For example, ERDA's budget for fiscal 1976 was $1.68 billion to research and develop new avenues of energy power for the United States. The breeder was allotted $430 million of this total, or 25 prcent of our overall energy budget—$110 million of it for the Clinch River breeder plant.

In 1972, the total program to develop the breeder was supposed to cost around $4 billion. By 1975, the federal government had already spent around $2 billion on the project—which was then slated to consume at least $10 billion of the taxpayers' money. The target is to construct and operate the breeder demonstration plant by 1983, which could pave the way for a slew of large-scale commercial breeders by the 1990s. The nuclear proponents blame the breeder's skyrocketing costs on inflation, delays, and safety problems. The breeder test facility, which Westinghouse is building for ERDA at Hanford, was supposed to be operating by

November 1977 at a cost of $530 million. In 1975 the target date was pushed back to late 1978, by which time the costs were estimated to run at least $622 million.

In 1973, the 350-megawatt breeder demonstration plant along Tennessee's Clinch River was budgeted at $699 million; by 1975 the figure had gone up to $1.7 billion, and was expected to climb beyond that—with ground for the plant yet to be broken because the Nuclear Regulatory Commission said it had to have more information about safety before it would consider ERDA's application for a construction permit. The administration, ERDA, and the nuclear industry, in their effort to justify the breeder, and drum up more enthusiasm for it in the Congress (which would have to fund it), keep pointing to the breeder program in Europe. France has two breeder reactor plants; England, two; and the Soviet Union, three. These plants are in the 200–350-megawatt range. The countries have said they want to build larger demonstration plants. West Germany and Japan have announced they will too. If this comes to pass, there could be ten 1200-megawatt breeder reactor plants worldwide by the late 1980s.

The liquid-metal fast-breeder reactor would use uranium-plutonium oxide fuel pellets surrounded by a blanket of uranium 238 placed around the core of the breeder. Some of the neutrons produced during the breeder fission process split the fissionable atoms of uranium 235 and plutonium to produce heat to create electricity; showers of neutrons released from the fuel rods go speeding out toward the blanket of uranium 238, which captures many neutrons to create plutonium 239, thus making it possible for the breeder to produce more potential nuclear fuel than it consumes. And 1 pound of plutonium, according to ERDA, has the energy equivalent of 3 million pounds of coal or 275,000 gallons of oil.

In January 1975, the soon-to-die Atomic Energy Commission disclosed that it had stockpiled around 200,000 short tons of relatively pure uranium 238.

This fertile nuclear material, the by-product of processing uranium for nuclear weapons and reactor fuel, can be used to produce enough plutonium for 2000 large 1000-megawatt liquid-metal breeder plants, according to the AEC, which said the rate at which this accumulated material could be used would depend on how much uranium 235 and plutonium 239 was available to fuel the proposed breeder reactor cores. The plutoniun 239 would have to come from fuel reprocessing plants, of which none were operating in the United States in 1975. And the uranium 235 would have to come from enrichment plants.

But more and more questions were popping up in Congress about plutonium and the breeder. And congressional opposition to nuclear fission power more than tripled from 1973 to 1975. Even so, efforts to postpone the development of the breeder in 1975 failed in the House by a 227–136 margin and in the Senate by 66–30. There has also been opposition from the U.S. Environmental Protection Agency, which in April 1975 suggested that construction of the Clinch River breeder should be delayed by 4–12 years. The Nuclear Regulatory Commission, although apparently convinced that plutonium 239 can be safely recycled today for use as a mixed-oxide fuel with uranium in contemporary light-water reactors, at first agreed to postpone its decision on the matter until 1977 or 1978.

Both of those motions were regarded as setbacks for the breeder. An EPA report said that less electricity would be consumed by Americans in the immediate future than the breeder advocates had estimated; therefore, EPA argued that the project could be delayed until its potential impact on the environment and the probable dangers of a major breeder reactor plant accident occurring could be more fully evaluated. EPA also said that the problems of storing plutonium waste should be discussed in greater detail.

The NRC slowdown came after agency chairman William A. Anders received a letter asking for a postponement from former Delaware Governor Russell W. Peterson, chairman of the President's Council on Environmental Quality, who said the NRC should wait until a special task force study on safeguards against the theft of plutonium delivered its report to the NRC in early 1976. "This threat is so grave," Peterson wrote in January 1975, "that it could determine the acceptance of plutonium recycle as a viable component of this nation's nuclear-power system."

Then the NRC reversed itself. On November 12, 1975, the agency said it was prepared to give "interim" licenses for facilities to extract plutonium from spent nuclear material and fabricate it for fuel. The critics said this indicated that the NRC would try to approve the plutonium recycle issue without the benefit of a full public airing of its hazards. And a suit to reverse the NRC's November ruling was filed the next month by six environmental groups, led by the Natural Resources Defense Council, and its attorney, J. Gustave Speth, who has been in the front ranks of the critics' fight against nuclear power.

ERDA seemed determined to push ahead with its research and development program on the breeder. On January 2, 1976, ERDA Administrator Robert C. Seamans, Jr., said the agency would make a final decision in 1986 on whether or not to promote widespread use of the breeder. ERDA seemed especially anxious to begin recycling plutonium, contending it could reduce the demand for uranium by 10 percent by the early 1980s. Unfortunately for the nuclear advocates, the earliest commercial recovery operation to reprocess plutonium from spent nuclear fuel is not expected to begin until 1977.

One of the people from NRC who I talked to said the lack of recovery plants was the most "crucial" link in the whole nuclear fuel chain, and that some atomic plants could be forced to shut down before 1977 because of this. There are several government fuel recovery plants, but these are not geared to process spent commercial fuel, from which 90 percent or more of the plutonium and uranium can theoretically be separated and decontaminated for future use.

On paper there are three commercial reprocessing

facilities in the United States—the Nuclear Fuel Services plant in West Valley, New York, about 30 miles from Buffalo; General Electric's $64 million Midwest Fuel Recovery Plant at Morris, Illinois; and the Allied-General plant in Barnwell, South Carolina. Fuel reprocessing plants are supposed to take in bundles of spent fuel rods, chop them into small pieces to expose the nuclear material they contain, then dissolve them in acid so that the reusable uranium and plutonium can be separated out for purification and eventual shipment to fuel fabrication facilities.

The West Valley plant, whose parent is Getty Oil, began operating in 1966; but it had to shut down in 1972 for repairs because radioactive leaks were contaminating workers and the environment. Nuclear Fuel Services, which hoped to start West Valley up again in 1978, continued to accept low-level wastes for burial; but it had to cut out this service in 1975 because the levels of radioactivity were rising in its burial trenches and the nearby creeks that fed into Lake Erie. In the meantime the plant was enlarged. It applied for a federal permit to triple its capacity to handle radioactive reactor waste. But the state of New York, on May 16, 1975, refused to certify the plant, contending that its radioactive discharges would exceed the state's water purity standards; and this raised a big doubt about the future of West Valley.

General Electric's plant was scheduled to begin operating in 1974. The nuclear industry was counting on the Morris plant. It was an experimental facility which was trying to simplify the recovery process and greatly reduce its costs. But the longest it could sustain one of its key steps was 26 hours, according to a GE report, which said the system kept clogging, and that repairing or replacing parts would be extremely difficult, costly, and dangerous because of the extremely high levels of radiation that would be present inside the heart of the plant. GE said it would have to put an additional $80–$130 million into the plant to make it work, and that the company would wait until 1978 to decide whether to go ahead with the Morris reprocessing facility or throw in the towel.

This leaves Allied-General Nuclear Services's $250 million plant in Barnwell, slated to begin operations in 1977, which an underfinanced group of intervenors in South Carolina has been trying to block.

Allied-General, a partnership of Allied Chemical, Gulf Oil, and Royal Dutch Shell, says it could make $75 million a year from the plant. But critics claim that allowing the plant to operate would result in radioactive discharges that would be too dangerous to tolerate. One of the radioactive elements that would be released from Barnwell is iodine 129, which has an extremely long half-life and is especially dangerous in terms of the human thyroid, where it can collect and build up to cause malignant tumors.

A report from the federal atomic laboratory in Oak Ridge, Tennessee, said iodine releases from Barnwell could be 50–100 times as great as Allied-General first anticipated. This, along with other radioactive elements released from Barnwell, could give people who live in the vicinity of the plant a much higher dose than government standards contend are acceptable. And some of this radiation from the reprocessing plant could collect in crops, milk, and fish that will be shipped to other areas in the United States. In addition to the everyday discharge of low-level radiation from the plant, Barnwell will collect huge quantities of radioactive waste from the spent reactor fuel it reprocesses that will be stored in large underground stainless steel tanks—with a 30–50-year lifetime—near the plant.

When Dr. John Gofman went to South Carolina in 1972 to testify against the Barnwell plant, he charged that if only 1 percent of this huge inventory (the accumulated radioactive poisons from 50 reactor plants a year) accidentally escaped it "could easily lead to damages in the neighborhood of $10 billion or more, to say nothing of the most massive civilian dislocations and suffering in human history." Gofman, who suggested this kind of catastrophe could be triggered by sabotage, an earthquake, or a tornado, said radioactivity on this scale would permanently poison 33,400 square miles of land. And Townsend Belser, Jr., of Columbia, South Carolina, now a lawyer but a former nuclear engineer who worked for Admiral Hyman Rickover, who is representing the local intervenors, complained that Barnwell is located in a class three, high-risk earthquake zone; but that the sitings of waste reprocessing plants are not governed by NRC regulations on earthquakes that apply to the siting of nuclear plants.

Allied-General, in a written statement presented to the AEC at one of the hearings on Barnwell in September 1974, said: "Unless this nation has adequate fuel reprocessing facilities, an important portion of the nuclear fuel cycle will choke and generating plants will be forced to shut down." The next month, L. Manning Muntzing, the AEC's director of regulation, called the lack of fuel reprocessing facilities "critical" in a memorandum circulated throughout the agency.

By January 1975, the same month its 29-year existence was terminated, the AEC reported that two nuclear plants in America would probably be forced to shut down before 1977 because there were no recovery plants to accept their spent fuel, and the plants' own on-site water-filled tanks were full of bundles of spent fuel rods. The AEC said that eight other reactor plants in America lacked enough space in their spent-fuel pools "to accommodate the discharge of a full fuel load in the event of an emergency." These spent fuel tanks at nuclear plants are massively built concrete chambers, open at the top, filled with about 40 feet of circulating water, which keeps the bundles of spent fuel from overheating, while a 10-foot cover of water provides the same protection from gamma rays emitted by radioactive krypton, iodine, cesium, zinc, cobalt (which colored the marbles at the Hanford Science Center), iron, and other radioactive elements—as a foot of lead or 4 feet of concrete. Most of these tanks at reactor power plants were designed to hold 1⅓–1⅖ cores of fuel bundles.

About a third or a fifth of the core—depending on whether it is a boiling-water or pressurized-water nuclear plant—has to be removed and replaced by bundles of fresh fuel after each operating year in order to get rid of radioactive by-products that build up inside the reactor to slow down the fission process.

Theoretically, these bundles of spent fuel are supposed to remain underwater in the plant pools for about six months. This is long enough for some of their radioactive heat to decay so that they can be moved out of the tanks and into huge double-walled casks, weighing around 20 tons, for shipment to storage facilities or fuel reprocessing plants.

A nuclear-plant pool with one or two years' worth of spent fuel in it can still accommodate a whole fuel core if a reactor has to be cleared to make repairs or because of an emergency; conversely, more than that in the pool would block a hot radioactive core from entering the tank. And even though a plant could operate without the capacity to use its fuel pool in an emergency, it would still have to shut down when the reactor had to discharge and reload, but couldn't because its tank was filled to capacity with bundles of spent fuel.

When nuclear museum director Lyle Wilhelmi and I left the Richland Science Center, we got a car from a government motor pool and drove north toward Hanford for an afternoon tour of the nuclear reservation.

In January 1943, when the federal government selected Hanford as the top secret site at which to produce plutonium for the first atom bomb, it was a remote, isolated, sparsely populated area surrounded by a vast expanse of desert. The Manhattan Engineer District Project, which was in charge of developing the atom bomb, took possession of 570 square miles along the Columbia River, relocated the 400 people living in Hanford, absorbed the tiny town, and proceeded to invest $350 million in the project. The government put up 4000 houses in nearby Richland for the Hanford workers, who labored feverishly around the clock to build nuclear reactors to produce plutonium and plants to process the poisonous metal. Twenty-seven months after construction began at Hanford, the world's first atomic bomb, made from Hanford-produced plutonium, exploded over the sands of New Mexico.

Hanford continued to churn out plutonium for the nuclear arms race. At one time there were nine plutonium production reactors along the Columbia River; by 1975 there was only one, which was also sending out electric power for the Washington Public Power System. This public utility was also building a $700 million, 1135-megawatt reactor power plant on land leased from the government at Hanford, which was expected to begin generating commercial electricity in 1978. The power company had initial construction permits to build four more reactor plants, two of them at Hanford, which could generate more momentum to turn Hanford into a bona fide nuclear park.

In fact Hanford, in some respects, already resembles a nuclear park. An Atlantic Richfield subsidiary, which operates the nuclear Science Center in Richland, is responsible for the storage of millions of gallons of high-level radioactive nuclear waste materials at Hanford. The firm also supplies plutonium for ERDA's liquid-metal fast-breeder reactor program. Westinghouse's subsidiary is building the fast-breeder test facility. Battelle Northwest has the Pacific Northwest Laboratories at Hanford, whose main thrust of study for ERDA is the effect of radiation upon the environment and health. And United Nuclear Industries, Inc., runs the government's dual purpose N-Reactor, which produces plutonium and commercial electric power.

Wilhelmi was trim, neat and precise, a middle-aged family man with flecks of gray in his dark hair. Like many people I've met who work for the nuclear establishment, he seemed a model citizen, active in community affairs, with an ardent overall interest in conservation and ecology. We drove north on the nuclear reservation, paralleling Rattlesnake Hill, the highest treeless mountain in the world. The sky was bright blue and the air was dry. Wilhelmi began to rhapsodize about the beauty of the desert—vast, expansive, and spiritually rejuvenating—which seemed to spur his insistence that nuclear power, with all its risks, was far less harmful to the environment than any other major source of contemporary power, even though the nuclear cycle is beset by so many problems, including what to do with the high-level, long-lived radioactive wastes at depositories like Hanford's.

In May 1975, for example, a federal task force study concluded that the future of atomic energy in the United States was seriously threatened by the uncertainties pertaining to the storage and disposal of radioactive reactor waste. The Environmental Protection Agency said a waste management plan proposed by ERDA was "inadequate." This involved storing waste in an aboveground "retrievable surface storage facility" —which critics charged would be too vulnerable to sabotage, war, or accidents.

ERDA apparently thinks a permanent solution to the problem might be achieved by burying radioactive waste in salt beds that have been geologically stable for 250 million years. "If this proves feasible, ERDA will allay the growing public fear that the benefits of nuclear power do not justify the risks," according to a *Business Week* editorial in March 1975. "And if it is not feasible," the magazine said, "the public has a right to know before committing itself to a technology that will be a perpetual threat to mankind."

Dr. Frank K. Pittman, a staunch pronuclear advocate and director of ERDA's division of waste management and storage, once put it this way: "The problem of waste storage is the least of any nuclear problem, but the problem is that you have to do it forever."

Dr. Alvin Weinberg, a physicist and the former director of the federal atomic facilities at Oak Ridge, Tennessee, another believer in nuclear power, touched on this point in December 1974, when he said: "No sensitive or even partly sensitive technologist today escapes this feeling of guilt, nor should we escape it. One of my colleagues asked me, 'What right does this generation have to make tons and tons of plutonium, probably the most poisonous material the world has ever known, without at the same time having a totally reliable and foolproof way of sequestering it forever, away from the biosphere?' This is one of those cosmic questions that bother one to death. The problem is that at the time we first unleashed nuclear energy, this was the last kind of question which occurred to us. . . . We are only human beings, and our crystal balls are dreadfully cloudy. . . . From now on [we should] make sure we have a consensus of the whole public. . . . The entire society can share the guilt."

The critics point out that countless civilizations have come and gone in a fraction of the time that this radioactive waste will have to be contained. John W. Simpson, the director of Westinghouse's reactor division and a past president of the American Nuclear Society, says, "I think that the worry about the ability of our institutions and social structures to last for hundreds of thousands of years is a needless one. If they do not last, civilization will not last and nuclear wastes will be one of the smallest problems."

At one time, the AEC thought it could permanently dispose of reactor waste by burying it deep in an abandoned salt mine near Lyons, Kansas. But the plan fell through because of local opposition to the scheme and the discovery that a nearby salt mine had leaks. Dr. Hannes Alfvén, a Nobel laureate in physics, addressed the salt-mine proposal by saying, "There is no doubt that the salt mines could be considered safe for any normal waste products. But because of the very large quantities of extremely poisonous substances, it is required that the repository should be absolutely free of leakage for a period of hundreds of thousands of years. No responsible geologist can guarantee this, simply because the problem is one of which we have no experience."

Other theories have been proposed to dispose of radioactive waste, among them orbiting it into space or burying it under mountains of glacial polar ice or deep below the floor of the ocean; but none of these have been developed because of the costs and uncertainties involved.

Congressman Mike McCormack, like many believers in nuclear fission power, doesn't seem to think that trying to get rid of reactor waste is a very serious problem. He is confident that scientists and engineers will be able to develop a technology in due time that will solve the problem. For the moment, he believes the best method is to solidify waste at a reprocessing plant into "a chemically inert 'glass.' "

"We'd seal up the glass in welded stainless steel canisters 1 foot in diameter and 10 feet high," McCormack told me. "Ten to 12 of these will hold all the waste a nuclear plant would produce in a year," he said. "The next step is to put a protective coat around them and bury them in stable salt mines. That could be permanent. Temporarily, we could put these containers inside concrete cylinders and stand them out in rows in the desert, where we could contain all the nation's nuclear waste to the year 2000 in a 2-mile-square area using this method."

Radioactive waste has been stored at Hanford since 1940s, and Hanford has been one of the chief depositories, ever since, for radioactive waste produced by the government. There are over 60 million gallons of intensely radioactive waste at Hanford, much of it stored in huge, underground steel-lined tanks encased in concrete, where the waste is artificially cooled or allowed to boil. Atlantic-Richfield, which handles the waste, divides it into three categories at Hanford:

• Low-level waste is solidified and buried.

• Intermediate level waste, extremely dangerous aboveground, goes into concrete covered trenches called cribs, which allow the radioactive material—including plutonium 239, cesium 137, strontium 90, and tritium (a heavy form of hydrogen)—to percolate down into the earth, where the most dangerous elements are supposed to stop well above the level of groundwater, which is hundreds of feet below the surface of the desert at Hanford.

• High-level waste is now stored in the underground tanks for several years, from which strontium and cesium will be chemically removed, leaving solid cakes of saltlike radioactive sludge at the bottom of the tanks.

Strontium and cesium, which comprise about 2 percent of the total radioactivity when the liquid waste is pumped into a storage tank, account for about 15 percent after three years and over 90 percent after 30 years. These two radioactive elements have to be contained for periods of time ranging from 600 to 1000 years. One 500,000-gallon tank at Hanford filled with radioactive waste could contain 200 times the amount of strontium 90 released by the Hiroshima bomb. Another way to view the toxicity of strontium is to note that federal standards warn that any amount over one billionth of a curie per gallon of water would be dangerous to drink. And by the year 2000 our nuclear reactors could produce billions of curies of strontium 90. The nuclear waste at Hanford represents a potential danger which would prove immense if anything happened to catapult it into the atmosphere where it could fall back to earth as radioactive rain.

Plutonium 239, because of its extreme toxicity, longevity, and explosive nature, carries its own special risks. One government report released in 1972 said enough plutonium had collected in the soil of the Hanford cribs to "conceive of conditions which could result in a nuclear chain reaction"—in this case, a criticality accident, or low-grade explosion. This led to the first plutonium mining operation in history to excavate and package the plutonium-contaminated soil for reprocessing and storage.

Hanford has also been troubled by leaking storage tanks: 11 out of 151 containers developed leaks, according to the AEC, between 1944 and 1970. The most celebrated incident occured from April to June 1973, when Atlantic-Richfield belatedly discovered that 115,000 gallons of potent radioactive wastes with an estimated 14,000 curies of strontium 90 and 40,000 curies of cesium 137 had leaked down into the ground. This happened not long after the American Nuclear Society said that "nuclear wastes have been successfully stored since the very beginning" of the atomic energy program, and that "massive tank failures resulting in large flows . . . have never occurred and are not expected." Monitors at the tank were recording the massive spillage, but the man in charge of the operation later said he had been too busy to notice that the huge container was leaking. The AEC later said the spill wasn't dangerous because all the radioactivity was contained underground, and insisted there was no way it could reach the atmosphere or groundwater.

But the critics were not put off; they said the massive leak indicated the role human error could play in the nuclear fuel chain; some claimed there was a danger that radioactivity could reach subterranean streams deep beneath the earth, and move in mysterious ways until it reached the Columbia River, some 10 miles away from the site of the spill. And the critics pointed out that abnormally high concentrations of accumulated radioactivity have already been found in water plants, fish, and wild fowl in the river.

One of the projects that Lyle Wilhelmi and I looked at on our tour of the Hanford nuclear complex was the great gray dome-shaped fast-flux test facility—120 feet high, 135 feet in diameter—which Westinghouse was building. This 400-megawatt liquid-sodium thermal fast-breeding reactor will not produce electricity; its sole purpose is to test uranium-plutonium oxide fuel mixtures, fuel rods, and assemblies; the economics of the breeder; and the ability of the reactor plant itself to hold up under exceedingly high operating temperatures and the flow of volatile, explosive liquid-sodium coolant.

Dr. Thomas B. Cochran, a physicist on the staff of the Natural Resources Defense Council who has written extensively on the breeder, warns that "it takes only a slight compaction of the plutonium fuel elements to trigger an explosive nuclear runaway." Some of the stainless steel fuel rods in an operating breeder could develop pinhole leaks, which in turn could make the rods corrode, blister, and swell. This could block the flow of sodium coolant, which in turn could cause the fuel rods to overheat, bulge, and fuse together. If enough plutonium collects together in one place, it could melt part of the core or make it go critical.

It would take about 22 pounds of plutonium oxide fuel—out of the many hundreds of pounds a large breeder would contain—to form a compact critical mass, according to atom-bomb experts. This could cause an explosion. It wouldn't be like a blast from a nuclear bomb; but enough energy might be released to rupture the reactor shell and containment building, which could permit tons of radioactive debris and explosive sodium coolant to escape.

It could happen very quickly—within less than a thousandth of a second, because the breeder, with its fast-moving neutrons, is smaller and more compact than the conventional light-water reactor. A sudden loss of coolant would permit the rest of the core to fuse into a molten blob that could ooze its way down through the foundation of the plant. Once this hot radioactive mass reaches the earth, and water there, a second, much larger explosion could occur.

Many nuclear critics say they are even more concerned about the vast amounts of plutonium that would have to be transported because of the breeder than they are about the dangers of a runaway reactor. Putting millions of pounds of plutonium "through the nuclear fuel cycle means that plutonium becomes a commonplace article of commerce," says Dr. John Gofman, "being handled by thousands of workers and being transported on highways, railways, and airways in numerous shipments per day."

A furor has already erupted over the discovery that planes carrying plutonium have flown into New York City's busy Kennedy Airport. Two 100-pound shipments of plutonium dioxide powder, authorized by the NRC, were flown from Brussels for delivery to the Westinghouse Fuel Development Laboratory in Cheswick, Pennsylvania. The first flight was in the summer of 1974; the second in February 1975. Each shipment contained 25 triple steel canisters of plutonium, which moved by truck to the Westinghouse facility after they left Kennedy. These canisters, according to the NRC, are designed to withstand a 30-foot fall, a blazing 30-

minute fire, or submersion under water for at least eight hours. But a fiery, explosive plane crash could be much more severe, and the plutonium from even one of these canisters dispersed into the environment would be enough, according to scientists, to trigger an epidemic of lung cancer and contaminate New York in particular and the East Coast in general with enough radioactivity to make the poisoned areas unlivable.

Then, right after the plutonium flights into Kennedy were disclosed, there was added concern when it was learned that the NRC had licensed pending shipments of British-made plutonium—290 pounds of it—into New York's international airport for future use in the Rochester Gas and Electric Company's Ginna reactor.

Congressman Les Aspin, Democrat from Wisconsin and a longtime critic of nuclear power, introduced a bill to ban the import of plutonium. "Plutonium should never be allowed to fly," Aspin told the *New York Times*. "It is too deadly a poison to spread around to be safe as air cargo." The *Times* itself, on March 29, 1975, used the word "incredible" in an editorial attacking the plutonium flights, complaining about "containers that would have broken into many pieces had the planes crashed," adding that the "revelation of these dangerous flights has brought justified anger in many quarters, as well as badly needed action by the new Nuclear Regulatory Commission."

The NRC staff, under mounting pressure, was temporarily ordered to stop processing licenses for the import of plutonium. "However plutonium is transported," the *Times* said, "it creates serious risks. If a ship carrying it should sink, unprecedented problems would be created, while trucks loaded with plutonium can crash or suffer other disasters that might liberate material into the atmosphere. Equally frightening is the prospect of the hijacking of plutonium for political or other purposes." This judgment indicates why transportation, which binds all the components or stages of the nuclear fuel cycle together, could turn out to be the most vulnerable link in the whole nuclear fission fuel chain because of the dangers that accidents, sabotage, and/or theft could occur when nuclear materials are shipped.

When Ralph Nader testified before the Senate Commerce Committee in June 1974, he said that if the nuclear establishment built and operated 1000 atomic power plants by the year 2000, we would have 60,000 yearly shipments of reactor fuel by truck, 10,000 by barge, plus 46,000 shipments of radioactive waste by truck and 10,000 by rail. This would add up to an annual total of 100 million nuclear miles a year. According to Nader, highway safety statistics conservatively report there's an average of one accident per every million traveled miles. Say we have 250 reactors operating in 1985. Nader contends we could then expect an accident a week transporting radioactive wastes around, and four accidents a week by the year 2000 if we have 1000 reactors around the country.

These shipments are likely to be far-flung; and each huge, thick lead and cylindrical steel cask—about 16 feet high, weighing 20 to 25 tons—would contain a formidable volume and level of radioactive poisons, moving past communities with their homes and schools and factories. Moving radioactive waste by train, considering the deteriorating condition of many miles of railroad tracks—on which serious accidents occur each year—poses special problems. At least one utility facing this danger wants to buy the right-of-way to 35 miles of railroad tracks for $1, fix them up, then lease them back to the railroad company. This is Philadelphia Electric Company, the principal owner and operator of the two large Peach Bottom nuclear-power units in southeast Pennsylvania just north of the Maryland line, which plans to ship its radioactive waste to the fuel reprocessing plant at Barnwell, South Carolina.

Philadelphia Electric wants to renovate the stretch of Maryland and Pennsylvania Railroad tracks that run from rural Peach Bottom to York, Pennsylvania, the county seat. People in the area, who were upset by the utility's proposal, hired York attorney Raymond L. Hovis, who has been fighting the Peach Bottom reactors, to oppose the scheme. Hovis contends the utility's radioactive shipments would impose a cancerous "burden" on the land along the tracks and to the people who live there because each of the large twin reactor units at Peach Bottom will send out a number of radioactive waste shipments a year.

Testimony at hearings on the Barnwell plant, for example, indicated that one large railroad shipment of radioactive waste could expose the people who live along the tracks that approach the massive plant to as much radiation in 17 hours as the federal authorities consider "safe" for the public to absorb in a whole year. And Allied-General, owner and operator of the Barnwell facility, said it expects to receive over 500 shipments by rail a year when the plant begins to operate. One of the first might come from Portland, Oregon, 2900 miles away. "It is conceivable," according to a company report on rail accidents in and out of Barnwell, "that such an accident might occur once every eight years." Also, unlike trucks, for which individual drivers are responsible, railroad cars full of heavy casks of radioactive materials may sit around for days in a switchyard, where no one would be specifically chosen to guard the car or cars from the danger of accident, sabotage, or radiation leakage.

There have been relatively few accidents thus far in the transportation of nuclear materials; and when they have occurred, most of the radioactivity has been contained, so far as anyone knows. One threatening accident occurred on December 20, 1973, when a truck carrying two casks of radioactive cobalt was involved in a multivehicle accident on a fog-shrouded, ice-covered highway near Stroudsburg, Pennsylvania.

Altogether there were six big trucks—one carrying flammable lacquer—and two cars involved in the pile-up. One of the cars burst into flames, the other car was also totaled, and five of the six trucks, including the one with the radioactive cobalt, were severely damaged, according to the state police report of the mishap. The man in charge of the cobalt shipment said, "If the truck had caught fire, it could have been very serious."

If a cask had opened, he said, people within 3 feet could only take a minute of exposure to the cobalt's beta and gamma rays, and people 20 feet from an open cask would show signs of radiation sickness after a half hour.

Another accident involved a New Jersey truck dock worker named Edward J. Gleason, Jr., who died after handling a leaking, unmarked box of liquid waste containing about 8 grams of plutonium 239. Shortly after the spill occurred in January 1963, Gleason complained of feeling ill. Four years later his hand and then his arm and shoulder were amputated because of a rare form of cancer, from which he died in 1973 at the age of 39. The leaking box began its journey at Hanford, had been near Pittsburgh for five years, and was on its way to the Brookhaven National Laboratory on Long Island when Gleason grabbed it at a Jersey City dock—which indicates the long distances, and probable points of transfer, that can take place in the shipments of radioactive material.

When Gleason tried to sue, company insurers fought his claim on the grounds that he could not prove a direct link between the spill and his cancer. According to reports, a sizable settlement was finally made with Gleason's widow.

The serious dangers of accidents thoughout the nuclear fuel cycle could be relatively slight compared to the risks society faces from the threat of nuclear theft and sabotage, according to reports prepared for the federal government.

In April 1974, for example, Senator Abraham Ribicoff, Democrat from Connecticut, released a special AEC study on nuclear safeguards which said this "danger is large and growing" and that the agency's standards were "entirely inadequate" to meet the threat of theft or sabotage. The report was written by Dr. David M. Rosenbaum, an outside consultant on terrorist activity and a former staff member of the White House Office of Emergency Planning. The five-man Rosenbaum study group, which included nuclear scientists and William Sullivan, former assistant director of the

FBI, concluded that "the potential harm to the public from the explosion of an illicitly made nuclear weapon is greater than that of any plausible power-plant accident, including one which involves a core meltdown and subsequent breach of containment. Acquisition of special nuclear material remains the only substantial problem facing groups which desire to have such weapons."

The safeguards report also expressed concern about the "widespread and increasing dissemination of precise and accurate instructions on how to make a nuclear weapon in your basement" and "the start of political kidnapping in the United States. It is our opinion that the kidnapping of Patricia Hearst does not represent an isolated and passing incident, but is rather the precursor of a wave of such incidents. If not firmly and completely met, these kidnappings may lead to a rise of urban terrorist groups in this country of a sort without precedent in our history. We believe these new factors necessitate an immediate and far-reaching change in the way we conduct our safeguards programs." The Rosenbaum report recommended the formation of a federal nuclear police force to protect atomic installations and the shipment of radioactive materials and a continuing strong liaison between the AEC, FBI, and CIA to protect plutonium and weapons-grade uranium.

Another study on the dangers of nuclear theft and sabotage, conducted by the MITRE Corporation, a think-tank organization in McLean, Virginia, seemed to echo many of the concerns raised by the Rosenbaum report. Disclosed in November 1975 by *Chicago Daily News* reporter Rob Warden, this second federal study, which also included Rosenbaum and Sullivan, said, "The acquisition of special nuclear materials by a terrorist group would give it a power of blackmail over the world at large . . . This is certainly the type of threat that should be taken seriously," adding that "perhaps licensed nuclear facilities [would be] a target."

The General Accounting Office, which released several reports in 1973 and 1974 on nuclear theft and sabotage, was also critical of the AEC's safeguards.

The GAO concluded that the nuclear industry's security measures would not prevent the theft of plutonium and uranium in transit and storage; and that a small group with weapons, perhaps as few as two or three people, could probably take over almost any of the nuclear-power plants in this country.

The Senate Committee on Government Operations, chaired by Abraham Ribicoff, put the safeguards issue in this perspective in 1975 when it said, "According to present projections, about 2 million pounds of plutonium will be generated each year by the world's civilian reactors by the year 2000. A system of safeguards will have to be devised to prevent the theft of as little as 20 pounds of plutonium, the minimum amount needed to fabricate a weapon." The committee also quoted testimony from Dr. Fred C. Ikle, director of the Arms Control and Disarmament Agency, who said terrorists could stage "an event that would change the world" by exploding a nuclear weapon in the heart of any major city.

Russell Peterson, the chairman of the President's Council on Environmental Quality in 1975, once suggested moving atomic plants out to sea, saying it was the only way he could imagine to guard them and population centers from "the potential of sabotage" in a growing age of terrorism—marked by bomb threats, hijackings of jet planes, and the kidnapping of diplomats and murder of Olympic athletes.

Terrorists in command of a reactor plant, or with a nuclear weapon in their hands, could use them as a threat to extort money, to demand the release of prisoners, to call for a change in government policy or for the resignation of public officials. These are all familiar demands for those of us who look at the six o'clock news, with its reports of well-armed, well-organized terrorist gangs operating throughout the world. Some of their names are familiar: Black September, Al Fatah, Tupamaros, Ulster Freedom Fighters, Symbionese Liberation Army, Black Liberation Army, and Japanese Red Army. According to a report from the AEC, there were over 400 incidents of terrorism during the six-year period ending December 31, 1973.

There have already been some nuclear threats. One of the most celebrated occurred in October 1970 in Orlando, Florida, when a blackmailer threatened to blow up the city unless he was paid $1 million in ransom and given safe escort out of the country. The demand was backed up by a convincing diagram of a nuclear bomb which a nearby air force armaments officer said would probably work. The ransom money was assembled, but police captured the "terrorist" before it was paid. He turned out to be a 14-year-old science honors student from a high school in Orlando.

In 1971 arson caused over $5 million in damages at one of Consolidated Edison's nuclear plants near New York City. Then, in November 1972, three armed men hijacked a jet airliner and circled over Tennessee, threatening to crash the plane into a nuclear installation at Oak Ridge unless their demands for $10 million in ransom were met. The hijackers failed to follow through with their threat; instead, they headed the plane for a crash landing in Cuba. Commenting on the incident, a spokesman for the AEC said a nuclear plant was built to withstand the impact of a 200,000-pound aircraft crashing at 150 miles per hour; but the critics retorted that a Boeing-747 weighs considerably more than that and travels much faster.

The following year guerrillas seized an atomic plant under construction in Argentina, decorated it with political slogans, then walked off with 15 weapons they grabbed from the plant.

In 1974, Commonwealth Edison's Zion nuclear plant received two bomb threats—one in June and another in August. And in May 1975, terrorists who said they were linked to anarchist groups in Germany and Spain took credit for two bomb explosions that ripped through a nuclear-power station under construction in France across the Rhine River from Germany, about 45 miles from Strasbourg.

But it's not necessary to plant a bomb in a plant,

or crash a plane, in order to destroy a nuclear reactor; it would be easier, and perhaps just as effective, according to the critics, to cripple the water main that carries cooling water into the reactor. A former underwater demolition expert for the U.S. Navy, testifying before a Senate subcommittee, said it would be relatively simple for a team of frogmen to attach underwater explosives to a nuclear plant's cooling-water intake.

The NRC, which succeeded the AEC in January 1975, has attempted to tighten up its security measures. But the Ford administration has been reluctant to increase the amount of money for nuclear safeguards. In 1974, for example, the White House cut $69 million from the program to guard against nuclear theft. As a result, the AEC was unable to hire 300 more guards the agency asked for, and was unable to begin developing a computerized inventory system and detection equipment for weapons-grade fissionable materials. Two armed guards, reporting to a checkpoint every two hours, followed by an escort vehicle with more guards, are now supposed to accompany every shipment of weapons-grade nuclear material that moves by truck or trailer—equipped with features to make them more theft-resistant; five guards on rail shipments will be mandatory; and two guards reporting to a supervisor must be on duty around the clock at reactor plants. These requirements, which seem modest to the critics, have been vigorously opposed by industry on the grounds that they have attached unnecessary costs and adverse publicity to the production of nuclear power.

Dr. Theodore B. Taylor, a nuclear weapons expert now active in the safeguards field, who says these new security measures could not turn back a "significant" armed attack mounted by terrorists, believes nuclear security could be upgraded by erecting formidable physical barriers at nuclear facilities, installing a network of sophisticated alarms and communications, and beefing up the number and caliber of armed nuclear guards.

Precautions advocated by Taylor include the shipment of radioactive materials in hard-to-hijack 100-ton containers equipped with devices that give off high-level radiation to ward off would-be thieves. All of this, according to Taylor, would only increase the cost of nuclear power in the United States by 1 percent. Taylor has told audiences that only 800 nuclear guards would be needed across the country by the year 2000 to protect the nuclear fuel cycle from sabotage and theft.

Mason Willrich, a professor of law at the University of Virginia and a prominent counsel on disarmament who has coauthored studies and a book on nuclear security with Theodore Taylor, also believes that "systems of national safeguards can be developed that will keep the risks of theft of materials for nuclear explosives from nuclear power industries at very low levels. Of course, 'zero risk' is an impossibility in this," Willrich adds, "as in other dangerous human activities."

In June 1975, the fear that nuclear security might be too lax was underscored by two former security men at Metropolitan Edison Company's nuclear-power plants near Harrisburg, Pennsylvania. Claiming they were fired because of their charges, the two men alleged that competing guard companies at the plant played "tricks" to discredit one another, which included unhooking gates to the plant; that over 300 unauthorized keys to the facility had been given out; that guards were insufficiently trained; security records were falsified; firearm tests for guards were rigged; electronic security devices did not work; and unauthorized people could walk in and out of the plant without much trouble.

In a prepared statement released through Ralph Nader's office, the two ex-guards, John Darcy and Joseph Shapiro, said in part, "We certainly do not regard our experience at Three Mile Island as an isolated problem. If lax security exists at Three Mile Island, it probably exists at other plants." Joe Shapiro later told the press that "it would be easy for a lone saboteur, as well as a group of saboteurs, to gain entrance to an operating reactor and hold this over the country's or area's head."

Dr. L. Douglas DeNike, a short, slight, bearded clinical psychologist from Los Angeles who has written extensively on nuclear safeguards for a number of publications, including the *Bulletin of Atomic Scientists,* has been working exhaustively without pay to publicize the dangers of nuclear power. DeNike once pulled the following stunt to demonstrate the inadequacy of reactor plant security, according to an interview he had with the Associated Press: In March 1974, DeNike went on a tour of the San Onofre nuclear plant near San Clemente, California. When he was in the visitors' viewing room, adjacent to the reactor control room, DeNike pulled out two objects from his pockets. One was a table knife labeled "lethal weapon." The other was an empty vitamin bottle labeled "nitroglycerin." "All the people I was with could have had hand grenades and pistols and not have been detected," DeNike told the AP.

One of the NRC ideas to reduce the publicized dangers of theft and sabotage is the nuclear park theory. The most advanced concept would cluster 10–40 reactor plants in an area of 25–80 square miles. These nuclear energy centers would also contain a number of facilities to enrich uranium, fabricate fuel, and reprocess and store radioactive waste.

Congressman Mike McCormack, who was enthusiastic about the idea, contended that nuclear parks would provide more control and better security against the threat of accidents, sabotage, or theft, especially in the shipment of nuclear material. Nuclear critics, for their part, literally shuddered at the thought of the nuclear parks envisioned by the likes of McCormack. The critics argue they might have a devastating effect on the environment.

A 40-unit nuclear park would require an immense amount of water to cool all its facilities—something like a million gallons a minute, double the amount consumed by a city the size of Chicago; and the heat released by this kind of center might provoke adverse environmental conditions. Wide swaths of land would have to be gobbled up to provide transmission corridors to transport electricity to consumers. The scheme would also require huge amounts of capital; and the cumulative, routine release of low-level radiation from the atomic installations could reach epidemic proportions in and downwind of the area, according to some scientists like John Gofman and Edward Martell, to say nothing of what might happen in the event of an enemy attack on a nuclear park or a successful attempt at sabotage.

Right now and into the foreseeable future, shipments of plutonium—with its atomic-bomb implications —are likely to be far-flung. Plutonium from the Barnwell plant, for example, converted from a liquid to a powder for shipment, would go out at least four fuel fabrication plants. Three of them are fairly close to Barnwell, but the one on the Hanford nuclear complex is thousands of miles away. All the plutonium that is fabricated into fuel would then be shipped to reactor plants, many of them hundreds or thousands of miles away from the facility where the fuel assemblies were made.

Theodore Taylor says that a crude atom bomb could be built from 9 pounds of pure plutonium 239 shaped into grapefruit size; or 17 pounds of plutonium produced by reactor plants—the kind that would be shipped from Barnwell; or 22 pounds of plutonium oxide for fuel pellets; or 13 pounds of uranium 233, produced by fissionable thorium 232 in a high-temperature, helium gas-cooled model of the breeder reactor; or 37 pounds of the highly enriched uranium used to fuel nuclear submarines and to power the high-temperature gas-cooled reactor plant at Fort St. Vrain near Denver. Testifying before a Senate subcommittee, Taylor said, "One person who possessed about 10 kilograms [22 pounds] of plutonium oxide could, within several weeks, design and build a crude fission bomb. By a crude fission bomb I mean one that would have a chance of exploding with a yield equivalent to at least 100 tons of high explosive, and small enough to be carried in an automobile. This could be done by using information that is easily accessible to the general public, and materials and equipment that are commercially available worldwide"—meaning a good encyclopedia and stuff that could be bought at any ordinary hardware store. According to Taylor, "This crude fission bomb could under some circumstances kill 100,000 people or more and destroy millions of dollars worth of property."

Taylor is a theoretical physicist who designed one of the smallest and lightest fission bombs—less than 50 pounds—when he was working for the AEC in Los Alamos. He also designed the largest-yield fission bomb ever exploded. In 1973 he took *New Yorker* writer John McPhee around the country on a tour of key nuclear installations. They visited atomic-power plants, fuel reprocessing centers, waste storage facilities, nuclear weapons sites, and shipping points. Taylor kept pointing out security inadequacies: gates were open, doors unlocked, fences low, accounting procedures casual; some guards were absent, others hadn't officially qualified with the pistols they carried. And he told McPhee about "weapons effects."

A nuclear bomb that would kill 5000 people if it exploded in an open field would kill 100,000 if it went off in a crowded football stadium—or in Manhattan's financial district; and just a sliver of the right kind of uranium "would be enough to knock down the World Trade Center in New York," Taylor said. Little Boy, the bomb that killed over 60,000 people and gave radiation sickness to thousands more when it fell on Hiroshima, was a 15-kiloton weapon. According to Taylor's conversations with McPhee, "A one-fiftieth kiloton yield coming out of a car on Pennsylvania Avenue would include enough radiation to kill anyone above the basement level in the White House. A 1-kiloton bomb exploded just outside the exclusion area during a state of the union message would kill everyone inside the Capitol. It's hard for me to think of a higher leverage target. . . . The bomb would destroy the heads of all branches of the United States government. . . . A fizzle-yield, low-efficiency, basically lousy fission bomb could do this."

The producers of a Public Broadcasting System show called "The Plutonium Connection," aired in March 1975, decided to put Taylor's warnings to the test, with the result that an unidentified 20-year-old chemistry major from Massachusetts Institute of Technology, using only published information, was able to design an atom bomb in five weeks. A nuclear physicist, evaluating the student's work, said the design could produce a bomb that might go off with the destructive force of 1000 tons of TNT.

No one seems to be sure how much plutonium and weapons-grade uranium which could produce one of these crude weapons is in the United States. But there might be a million pounds or more of it in storage around the country. Trying to keep track of it has been difficult, and might be impossible if the breeder reactor springs into being, sending millions and millions of pounds of plutonium moving through the nuclear fuel cycle. Some of this material is already missing. The government calls this MUF—materials unaccounted for. It could be stolen or embedded in machinery. Facilities that handle 2½ pounds or more of weapons-grade nuclear material have to be licensed by the NRC.

One of the most celebrated of the MUF cases occurred in the fall of 1965, when a Nuclear Materials and Engineering Corporation plant at Apollo, Pennsylvania, came up 207 pounds short in its inventory of highly enriched uranium, worth about $1 million at the time. The AEC had the plant closed down and began looking for the material. Thirteen pounds turned up in air filters at the plant. Another 15 pounds was discovered at a mountaintop burial pit, 8 miles from the plant. But there were still 148 pounds missing at the end of the search, and the company was finally forced to pay the AEC $834,000—the cost of the missing uranium.

Weapons-grade uranium was selling for around $15,000 a pound in 1974; the price of plutonium 239 per pound was around $5000—five times as much as heroin and ten times the cost of gold. We can only guess how much these fissionable materials might be worth on the black market. These are the kind of prices that might make uranium and plutonium an irresistible lure for criminals and thieves.

As early as 1969, AEC Commissioner Clarence E. Larson was worrying that "once special nuclear material is successfully stolen in small and possibly economically acceptable quantities, a supply-stimulated market for such materials is bound to develop. . . . As the market grows, the number of thefts can be expected to grow. . . . I fear such growth would be extremely rapid once it begins. Such a theft would quickly lead to serious economic burdens to the industry and a threat to national security."

The critics point out that 2 percent of everything shipped today in this country is pilfered. They also repeat allegations that organized crime has made deep inroads into the transportation industry. Internal crime, often committed by employees, reportedly costs business in the United States $15–$30 billion a year, according to educated estimates in 1975. And the U.S. Army's Physical Security Review Board, in a report made public in September 1975, said that "organized criminal elements" have already taken advantage of lax security to steal thousands of American military weapons and explosives in the United States and abroad. The army investigators said some of these arms were sold to criminals and terrorists for "huge profits," indicating that a precedent has already been established that might lead to the theft of nuclear materials on a wholesale level.

In an apparent attempt to minimize the danger, L. Manning Muntzing, then the AEC's director of regulation, wrote in October 1974 that "it is believed that the problem of safeguarding plutonium is not as difficult as the problem of controlling heroin." This flabbergasted the nuclear critics, who cried that the criminal flow of heroin seemed impossible to stop; that plutonium was already five times as expensive as heroin, and would in time be hundreds of times as commonplace, and hundreds of thousands of times as dangerous as anyone would ever imagine heroin could be.

Trying to keep all this plutonium under the lid might be impossible; any meaningful attempt to safeguard plutonium could turn the country into a garrison state, because there would always be people involved. Congress heard in 1974, for example, that more than 3600 persons with access to nuclear weapons had been removed from their jobs in a year's time. Some of them were considered security risks because of alleged drug abuse, mental illness, alcoholism, or problems of discipline. One of them was a top national security officer for the AEC. Reports identified him as a compulsive gambler who had borrowed $239,000 from his fellow employees at the AEC from April 1964 to June 1972, of which he failed to repay over $170,000. The fact that he had access to atomic secrets at the same time his superiors were unaware of his fondness for the racetrack, could have made him a prime target for blackmail.

Ordinary citizens could also wind up in a nuclear dragnet of the future. State police in Texas have already collected dossiers on opponents of nuclear-power plants. One of them was Robert W. Pomeroy, a commercial airline pilot based in Texas and the chairman of a local antinuclear group. He was put on the police "subversive" list after appearing before a city council meeting to protest the construction of a nuclear-power plant. Pomeroy's file, released after a lawsuit, said, "Sources feel the subject is using Citizens Association for Sound Energy as a front group, possibly for a Ralph Nader action."

Ralph Nader himself says, "Some observers believe there will be a million people with direct and backup assignments to guard the nuclear-power industry by the year 2000. Dossiers and surveillance on hundreds of thousands of people will proliferate along with elaborate security clearances, lie detector tests, and invasions of privacy. There are no clear limits to such a garrison mentality, once it gets under way, except one: we don't need nuclear power in this country."

David Comey, nuclear critic from Chicago, puts it this way: "We have never gone to a national police force and widespread clandestine surveillance in the past because the hazards that potential 'subversives' and saboteurs could exploit would not justify the effort; but in a plutonium economy the hazards are such that such strict safeguards may be demanded. In fact I can foresee the situation developing to the point where the single most important function of government will be the exercise of special control over nuclear material; when the dominant obligation of government will be protecting radioactive waste materials from a potential enemy missile strike, subversives, and terrorist groups. And this would require the formation of a national police force.

"We may wind up with a radioactive waste czar who will become more important than the Joint Chief of Staffs are today," according to Comey. "Today the ultimate power may lie in the command of the 'little black box' controlling the deployment of the Strategic Air Command's nuclear warheads. Tomorrow, in 1984, the ultimate power may be the key to the storage facility for radioactive waste materials."

Comey also worries that terrorists "could easily make a radiation device using no technical devices whatsoever. All they would need is a small amount of plutonium shavings; seal them in an ordinary plastic chop-suey container; tape it to the end of a magnesium flare like those the truckers use; take it all up to the top of the nearest tall city building and—let it burn and go down. It could give lung cancer to 50,000 people downwind."

The nuclear danger to our civil liberties was dramatically raised in 1975 by Harvard law student Russell W. Ayres, who startled an antinuclear audience by posing this hypothetical situation: A nuclear-bomb threat has been made. The bomb is ready to go off. The police have the terrorist in their custody. "Should the police torture the terrorist" in an effort to save the lives of thousands of potential victims? Ayres asked.

He pointed out that "plutonium provides the first rational justification for widespread intelligence gathering against the civilian population. In the past, federal courts have taken a skeptical view of attempts to justify spying on national security grounds, but with the very real threat of nuclear terrorism in the picture, the justification is going to sound very convincing."

Ayres underscored all this with his reminder that "to the extent that we have civil liberties at all today, it is because we have not had to ask questions like whether it is better to torture a suspected terrorist than to let a city go up in flames."

There is also increasing concern, in virtually all quarters, that the spread of reactor plants around the world might also spread nuclear weapons around as a result. Take the case of India, for example. It had a Canadian reactor, and was able to divert enough plutonium from the reactor to make an atom bomb, which it exploded in May 1974. This was extremely disturbing—a dire indication that any nation with a reactor could come up with enough nuclear material for an atom bomb.

To put this in some perspective, there were about 120 nuclear reactors in 23 countries in 1975. Of these, only six nations had the bomb for sure; the United States, France, Britain, China, the Soviet Union—and India. But there could be others. Israel probably is one. There has only been one major attempt to limit the spread of nuclear weapons: the Nuclear Nonproliferation Treaty, signed by 84 nations in 1968. It prohibited the nuclear club from giving weapons away, or assisting other nations in their development; the nonnuclear nations promised they would neither accept nor manufacture nuclear weapons.

Unfortunately, China, France, and India did not sign the treaty; and a dozen other countries that have expressed interest in nuclear weapons have either refused to sign or ratify the agreement—including Israel, Egypt, Japan, Argentina, Brazil, Taiwan, Libya, and South Korea. Some of these have definite nuclear weapon ambitions. West Germany signed a multibillion-dollar contract in 1975 with Brazil to provide the South American country with several large reactor plants, fuel reprocessing facilities, and a uranium enrichment plant —which could produce weapons-grade material.

Walter F. Hahn, a foreign policy expert who has written widely on the nuclear subject, put this situation into troublesome perspective in 1975 when he wrote: "The world is populated by literally thousands of scientists and engineers who have the know-how to fashion a rudimentary atomic device. The major stumbling block to nuclear ambitions is thus not scientific and engineering skill, but the access to the needed fissionable material in the form of plutonium. This fissionable material, however, will become increasingly available as a by-product of the peaceful harnessing of nuclear energy. By 1982 the world's nuclear-power stations are expected to produce some 100,000 kilograms [220,000 pounds] of plutonium a year—enough to fabricate tens of thousands of nuclear weapons."

A Vienna-based organization called the International Atomic Energy Agency is trying to police the flow of enriched uranium and plutonium in the world; but it is woefully understaffed and its jurisdiction is limited to those countries which permit IAEA to operate within their borders.

In an effort to increase America's role in the international nuclear safeguards field, some senators have introduced legislation to control the export of American-made reactors, materials, and other nuclear hardware. Senator Abraham Ribicoff, staunchly pushing this bill, said, "There is an urgent need to require entire fuel cycle safeguards as a condition of nuclear sales." Senator Charles H. Percy, Republican from Illinois, another sponsor of the measure, in April 1974 said, "We are literally racing the clock. The day when the United States held a monopoly on nuclear technology is long past. As a practical matter, we have a very short time—perhaps two or three years . . . during which we can lead the world toward adequate safeguards."

As I was driving through the desert at Hanford, listening to nuclear museum director Lyle Wilhelmi extolling the virtues of the "peaceful atom," I couldn't help but brood about the many problems in the troubled nuclear fuel cycle. And suddenly I found myself thinking of the physicist Hannes Alfvén. At one time he has been a true believer in nuclear power; then he became disillusioned. Alfvén felt that the dangers of the nuclear fuel cycle were insurmountable, and that in the end everything that had been done to control them was, and would continue to be, "irrelevant" and "pathetic." Alfvén summed it up this way: "Fission energy is safe only if a number of critical devices work as they should, if a number of people in key positions follow all their instructions, if there is no sabotage, no hijacking of the transports, if no reactor fuel processing plant or reprocessing plant or repository anywhere in the world is situated in a region of riots or guerrilla activity, and no revolution or war—even a conventional one—takes place in these regions. The enormous quantities of extremely dangerous material must not get into the hands of the ignorant people or desperadoes. No acts of God can be permitted."

VII
The Dollar Goes Up

On the morning of September 10, 1974, 39-year-old nuclear critic David Comey marched into one of the grand conference rooms atop the Conrad Hilton Hotel on Chicago's famed Michigan Avenue to deliver a devastating attack on the economics of nuclear fission power plants. Comey's well-documented thesis went like this: Reactor plants are unreliable and highly inefficient; they produce less and less electricity with increasing age, because the older they get, the more they break down; and this adds billions of dollars to the cost of nuclear-generated electricity.

The occasion was one of the Project Independence hearings that the Federal Energy Administration was holding around the country in 1974. In theory, the sessions were supposed to gather information to feed into the makings of a national strategy, designed to make the United States an energy self-sufficient nation by 1980; in theory, attention was paid to the benefits of conserving energy; reducing the environmental impact of mining and burning coal; stepping up the development of new sources of power such as solar and geothermal energy; but in actuality, the four days of the hearings in Chicago evolved into an ongoing debate over the merits and shortcomings of nuclear power.

In fact John C. Sawhill, then chief of the FEA, set the tone of the meeting in his opening address by declaring, "We're here today to discuss nuclear energy because of the potential it has to become a major energy source of the nation." He went on to say that "the weight of the evidence suggests that nuclear power is among the safest sources of energy we have on hand today"—an argument that was attacked time and time again at the conference. Then Sawhill made a passing reference to the fact that "construction costs" of atomic-power plants "are high," and that "the nuclear industry needs to improve the reliability of operating plants: equipment failures have been a major problem at new plants."

This set the stage for Comey's documented testimony. As Comey later explained, "I began to get interested in the economics of nuclear power when I was asked to testify before the FEA. . . . I knew the FEA would be indifferent to the safety hazards of nuclear reactors and to the moral and social problems of the nuclear fuel cycle. So I decided to focus on something their crabbed mentalities could comprehend: money."

The environmental director of the Chicago-based Business and Professional People for the Public Interest charged that the public might have to pay for a cost overrun of $100 billion or more over the next 15 years, because nuclear plants weren't nearly as efficient as they were touted. Comey, as usual, had the data to back up his charges; and what he had to say at the Project Independence hearings made headlines around the world, touching off a furious debate that has been going on ever since about the economics of nuclear plants.

Prior to Comey's testimony—in fact, that same morning—the nuclear establishment's credo, as always, was that the power it produces is safe, clean, efficient, and cheap. Nuclear plants had been sold, financed, and built on the assumption that they would generate electricity at 80 percent of capacity or more—compared to 65 percent for fossil-fuel plants, which are cheaper to build—after a nuclear plant had three or four years to iron out the breaking-in kinks.

However, in 1973 the Atomic Energy Commission decided to make a study of nuclear-power-plant efficiency for the reactors operating that year. This AEC report was available in 1974, for those who knew where to look. Comey got his hands on it. And this is what he found: Nuclear reactor units licensed to operate for 30–40 years were operating at an average of only 54 percent of capacity; the older they got, the worse their performance—down to an average 38 percent of capacity for plants more than six years old.

There are several reasons for this lag in capacity, among them the fact that the regulatory agencies won't permit some reactors to operate at full blast because they might have a catastrophic accident; also, nuclear plants aren't particularly reliable because they're prone to breakdowns, which often take an exceedingly long time to repair—and this can be very expensive.

Comey said that this gap between the maximum capacity of reactors to produce and their actual output could add $121 billion to the nation's power bill by 1990 if nuclear plants generate at 54 percent efficiency—and more than $198 billion if they produce at only 38 percent of capacity.

After noting that reactor efficiency reaches a peak during its three- to four-year breaking-in period but then begins to drop away after the fifth operating year, Comey said: "Corrosion problems set in, leaking fuel becomes a problem, and system components break down due to fatigue and other wear-related problems. An additional hazard is the accumulation of highly radioactive crud in the primary system, which means that any repair work on this system will consume enormous amounts of time and personnel in order to avoid excessive radiation exposure. In some instances, thousands of workers have had to participate in the repair of a single plant, and a worker can receive his maximum permissible [radiation] exposure after working on the primary system for less than 60 seconds, thus 'burning him out' for the next three months."

Comey later pointed out that 350 men were required to make prolonged repairs at one Commonwealth Edison reactor plant—work that 12 men could have quickly completed at a conventional fossil-fuel power plant. This could be one of the reasons why a number of utilities are having second thoughts about building nuclear plants. Louis H. Roddis, for instance, a former vice-president of Consolidated Edison Company in New York and head of its nuclear division and a past president of the industry-oriented Atomic Industrial Forum, caused a considerable stir by describing what happened when a cooling pipe had to be repaired at one of Con Ed's nuclear plants in May 1970.

"The total effort to locate the failure, analyze the causes, design the repair, and make the repair took seven months," Roddis said. "In the seven-month effort, in order not to exceed radiation-exposure regulations, 700 men were used. The job required, at one time or another, the use of every welder in the Con Ed organization who was qualified in a certain welding technique. . . . A similar job on a conventional plant would have required two weeks and would not have involved more than 25 men."

Consolidated Edison, the nation's largest public utility, has had plenty of troubles with its nuclear-power stations. Its 265-megawatt Indian Point Unit 1 on the Hudson River above New York City, for example, which began generating in the 1960s, was out of service all year in 1973 and again in 1975. Indian Point Unit 2, an 873-megawatt reactor, started up in 1973 but operated on an on-and-off again basis because of breakdowns and repairs. Then there was a continuing string of problems—accidents, breakdowns, and delays —during the construction of Indian Point Unit 3, a 965-megawatt nuclear-power plant.

This reportedly contributed to the derating of Consolidated Edison's bond issues to finance the plant. Interest rates went up. The utility, trapped by rising interest and operating costs, failed to pay a dividend in the first quarter of 1974—virtually unheard-of and for many, a frightening thing for a big electrical utility company to do. Wall Street was noticeably glum, and some officials in Washington, D.C., were worried.

Consolidated Edison, in order to bail itself out, got the state of New York to buy its new Indian Point reactor plant and a fossil-fuel plant in Queens; and the company, of course, asked for a rate increase. The costs of electricity have been going up everywhere, triggering consumer protests, but New Yorkers are paying the heaviest dues; their bills increased by more than 37 percent from 1973 to 1974—the biggest hike in the nation, according to the U.S. Department of Labor. Many observers blamed the increase in part on Consolidated Edison's troubles with its nuclear fission power reactor plants.

When Comey began talking about billion-dollar-plus gaps between the capacity to produce electricity and the actual output of nuclear plants, he began to touch on another sore point of the nuclear establishment—the increasing difficulty of attracting enough money to build new plants.

"The annual requirements for new capital for the entire electric utility industry have recently bounced up from $5 billion to roughly $10 billion, and the industry is having grave difficulties raising even these sums. It is unlikely," Comey continued, "that the investment banking community will be receptive to cost overruns of the magnitude described above. Since only the U.S. government is likely to be available for these capital requirements, it may very well be that the names of General Electric, Westinghouse, Combustion Engineering, Babcock and Wilcox, Consolidated Edison, Consumers Power, and others will join Lockheed, Boeing, and Grumman on the rolls of corporations bailed out of costly technological misadventures by the taxpayers."

Prior to Comey's testimony, Thomas G. Ayers, chairman of Commonwealth Edison Company of Chicago, the country's biggest nuclear utility, told the hearing that Edison's reactor plants were available to generate electricity 75 percent of the time, compared to 67 percent for its coal-burning units. This made nuclear energy more reliable than other sources of power, according to Ayers, who said Edison's reactor plants supplied Chicago and northern Illinois with about 30 percent of its power in 1973. A few months later, in an interview, Ayers told the press that it cost Commonwealth Edison 22 percent less to generate electricity from nuclear plants than from any other source.

Comey claims the proportion of time that reactor plants are able to produce is of far less importance than how much electricity they actually generate over a given period; output determines the profit for utilities and service and bills for consumers. Figures published by Comey in 1975 for the previous year, again based on government reports, indicated that Commonwealth Edison's seven operating nuclear plants were producing at an average of 39.5 percent of capacity in 1974, which prompted Comey to say: "If their record turns out to be a norm for the industry, then twice as many nuclear plants will have to be built to produce the same amount of electricity as would have been produced had the plants operated at the 80 percent of capacity factor on which the AEC's cost-benefit calculations were predicated. This means twice as many capital dollars expended, not to speak of additional operating costs. I see no difference between this situation and a weapons system that only produces half the 'bang for the buck' originally projected by the Pentagon. We call the latter 'cost overruns'; I see no reason not to apply the same terminology here."

Comey also noted in 1975 that the average electrical output from nuclear plants around the country in 1974 was about 52 percent. And something perhaps even more interesting: 800-megawatt nuclear plants and larger included in the 1974 AEC study were producing at an average 46 percent of capacity, which meant that the overall reactor plant productivity average was being buoyed up by units that were smaller than 800 megawatts. "This," according to Comey, "has implications for the future," because most of the reactors "currently under construction are larger than 800 megawatts."

In March 1975, a report made public by the NRC concluded that the utilities in this country have not been concerned enough about the safety and reliability of their reactor plants. The study also said that public utility commissions on the state level "have little or no influence" on the design of reactors that could make them more productive. Incidentally, Dr. Edwin G.

Triner, director of policy planning for the NRC, who headed the study, used a slightly different group of reactors than those used in the report analyzed by David Comey to conclude that average reactor capacity was a bit under 54 percent.

The data gathered by Dr. Triner indicated "there is no evidence" that the utilities "have contractually imposed reliability standards upon their architect-engineers" to improve reactor reliability and output. Utilities were also loath to incur the extra expenses that improving the design of a reactor plant might entail, according to Triner; he added that the architect-engineers who design and build reactor plants have a "short-term" interest in the units rather than an ongoing commitment to their reliability.

Dr. Norman Rasmussen, professor of nuclear engineering at MIT, told the Atomic Industrial Forum in 1974 that "one of the most serious issues that the intervenors can raise today, with good statistics to back their case, is that nuclear plants have not performed with the degree of reliability we would expect from machines built with the care and attention to safety that has often been claimed for nuclear plants."

William E. Heronemus, professor of civil engineering at the University of Massachusetts, who supervised the construction of nuclear submarines for Admiral Hyman Rickover, says it would cost much more to make nuclear plants as safe as nuclear submarines. Testifying at a licensing hearing for a nuclear plant in 1973, Heronemus said it cost $2400 per kilowatt to build one of the navy's nuclear submarines, compared to $400 a kilowatt for the plant under discussion; and that he would have refused to pass wiring and piping for the navy that had been accepted and incorporated into the nuclear plant.

But nuclear-power plants are crushed between the demands of quality and costs in a way that the nuclear submarine program was not. Even though Comey and others have warned that large nuclear plants may be even less reliable than smaller units, the utilities want to build ever larger plants—and two or more reactor units at the same time, on the same site—in an effort to hold down the capital costs of nuclear plants, under the assumption that this will improve the economics of nuclear power. Costs have been skyrocketing because of inflation, the rising price of labor and materials, and the added expenses that have been tacked on to nuclear plants in an effort to make them safer—mostly in response to what the antinuclear critics have learned, protested, and publicized. A nuclear plant that cost $200 million to build in 1968 was priced at $1 billion by 1975. For example, when Philadelphia Electric Company began to talk in the early 1970s about its plans to build two 1140-megawatt, high-temperature, gas-cooled reactor units along the Susquehanna River across from Peach Bottom, the utility estimated the Fulton plant would cost about $1.4 billion. In 1975 the utility said it would cost $2.5 billion. As Anthony Bournia, the Nuclear Regulatory Commission's project supervisor for the Fulton plant, sourly told me, "We used to talk about $200 a kilowatt to build a reactor plant; now it's up to $1000 per kilowatt."

From the 1950s on it was commonplace for the nuclear establishment to say that atomic power would be much cheaper than fossil-fuel-fired plants. This was based on the assumption that the cost of uranium to power nuclear plants would be negligible compared to the price of oil, gas, or coal. Back in the 1950s, when Wall Street investment banker Lewis Strauss was chairman of the AEC, he assumed that nuclear power was going to be so cheap that it wouldn't have to be metered. A spokesman for Chase Manhattan Bank put it this way in 1957: "Conventional power costs will creep up with the slow but steady increase in the cost of fossil fuels, whereas nuclear energy will come down in cost."

The price of diminishing supplies of oil and gas has been going up and up; so has the price of coal. But the price of uranium has been skyrocketing too.

In 1975 the Investors Responsibility Research Center, a Washington-based investment advisory service, calculated that it would cost $811 milltion to begin building a 100-megawatt nuclear plant in early 1975, compared to $638 million for a coal-burning unit of the same size.

Westinghouse, saying it was taking into account capital investment, the price of fuel, operation, and maintenance, in early 1975 claimed that it cost 42.6 mills per kilowatt-hour to produce electricity using an oil plant in 1975, 33.2 using coal, and only 24.7 using nuclear fission. This, according to the reactor manufacturer, translated into annual savings—via the nuclear route—of $64 million over coal and $135 million compared to oil-fired plants.

This is an issue that is fogged by many difficult-to-balance factors, compounded by lack of statistical data. But Comey and Westinghouse agree that nuclear plants had to produce at 55 percent of capacity or better in 1975 in order to generate cheaper electricity than coal-burning power plants. And economists working for Friends of the Earth, the San Francisco-based environmental group, used the Investors Responsibility estimates in 1975 to come up with this: operating at 75 percent of capacity, nuclear power costs about 18.5 mills per kilowatt-hour compared to 14.6 mills from a coal plant; and the cost per kilowatt-hour for nuclear plants, producing at 55 percent of capacity, is 25.2 mills a kilowatt-hour.

"These predictions," according to David Comey, "were borne out by the 19th Steam Station Cost Survey made by *Electrical World,* the utility industry newsweekly. Their survey, which appeared November 15, 1975, showed that electricity from nuclear plants actually averaged 18.2 mills per kilowatt-hour, whereas electricity from coal-fired plants cost only 13.5 mills per kilowatt-hour."

The Ford administration, which gave nuclear fission almost half of the energy research and development money for fiscal 1976, contends that nuclear power is our best energy option. As Frank Zarb, head of the Federal Energy Administration, said in July 1975, "Even if no improvement were made in nuclear-plant productivity, nuclear power would still be a bargain for the consumer," adding that the United States would be unable to produce the electricity we need unless we push ahead with the development of nuclear power.

Robert I. Smith of the Edison Electric Institute, an industry association with offices in New York and Washington, told a House subcommitte that a 1000-megawatt reactor plant saves up to a million barrels of oil per month. "At a nominal $10 per barrel for oil, this would save $10 million per month from flowing overseas," according to Smith. But Comey says a "1000-megawatt nuclear plant would have to operate at an average capacity factor of 93.3 percent to save this much oil."

In 1975 President Ford was calling for the construction of 200 new nuclear plants by 1985. But rising costs, the difficulty in attracting capital, time-consuming and expensive bouts with antinuclear critics, opposition to rate hikes, the general uncertainty about the future demand for energy, and the future of nuclear power in particular—all have cut into the construction of nuclear plants.

According to figures released in early 1975 by the NRC, 126 of the 180 nuclear plants that were planned or under construction in 1974 were deferred; 14 units were canceled outright. Dr. Edwin Triner, the NRC planning director, said in an internal office memo that the utilities decided to hold off on reactor units because of the time it takes to build an atomic-power plant—eight to ten years, coupled with the uncertainties about the demand for power. Another Triner memo said, "The utility people have seen lower sales and peaks. . . . They therefore conclude that all the extra capacity previously planned will not be needed and should not be built." As Triner put it, "For many reasons, the utilities find it to be in their interest to play a waiting game. They don't know what their demand will be."

Power plants have been built on the assumption that the demand for electricity would increase steadily at least 7 percent per year. In 1974, according to the Edison Electric Institute, this trend came to an abrupt halt as consumers began to hold back on the consumption of electricity. It was the first time this had happened since 1945, the institute said, noting that this had occurred before only during the depression years of 1932 and 1938. This holding trend continued into 1975, according to the Institute, which said that demand for electricity, when compared to the same 29-week period in 1974, went up only 1.4 percent. This seems to bear out the claim of critics like Dr. Arthur Tamplin, biophysicist now at the Natural Resources Defense Council in Washington, D.C., who contends that conservation will be our "greatest energy resource in the coming years."

David Comey is one of many who says we could further reduce the demand for electricity by time-of-day pricing and making industry pay more for power. This, according to Comey, would "flatten out the peaks, which are more than twice the average demand. They only last a few hours around the middle of the day during the summer—around 1:00 p.m., and from 3:30 to 6:30 in the afternoon during the winter." Initiating time of-day pricing, according to Comey—as the telephone companies do on long-distance calls, along with boosting the bills for industry, would encourage heavy consumers of electricity to reschedule certain of their activities to the most economical periods of the day or night.

The nuclear industries and utilities began to act as if they were under siege. As John W. Simpson, director of Westinghouse's reactor division, said in 1975, "A problem that does pose difficulty is the financing of a nuclear-based electricity generating system. A program on the scale now under discussion (to make the United States dependent upon nuclear power for 50 percent or more of our electricity by the turn of the century), which includes building all generating plants and transmission lines, will require a capital investment of about $600 billion in 1980 dollars between now and 1990. This is an amount even greater than the national debt and is four times the present net investment in plant facilities."

William R. Gould, chairman of the Atomic Industrial Forum in 1974, and executive vice-president of Southern California Edison Company, insisted that "it is quite clear that prompt and adequate rate relief is the step that would do the most good in the shortest time."

But ratepayers were already screaming about their high electric bills. The National Utility Service says a survey of the country's 24 largest utilities showed residential rates rose 37.5 percent from June 1973 to December 1974. No other living costs rose that fast. The Labor Department's consumer price index said the cost of power went up by 38.8 percent after March 1973. Only food, which rose 36.8 percent, was close to the hike in electricity. New Yorkers, hit the hardest, were followed by consumers in Boston with a 32.8 percent increase and Philadelphia with 26.4 percent. These rate increases sparked a consumer revolt around the country. When Philadelphia Electric Company—which has large interests in seven reactor plants, including some which may never be built—went before the Pennsylvania Public Utility Commission in September 1974 to ask for another rate increase—this time for $112 million, the hearing was suspended because of the uproar created by angry consumers who jammed the meeting. Some of them were armed with a National Utilities Services report alleging that the average Philadelphia Electric customer was paying 63 percent more for electricity during the first six months of 1974 than for the same period the year before.

Philadelphia Electric, the sixth largest utility in the nation—which in 1973 said, "We are planning for 70 percent of our electric generation to come from nuclear plants by the mid-1980s," has been forced to back-pedal. In a letter mailed in 1975 to company managers and supervisors, the utility called for "belt tightening and austerity . . . made necessary by changing economic, social, and political conditions," adding that "new generating capacity needed to serve customers through the year 1990 will be 43 percent less than estimates made as recently as April 1974. This will greatly affect our generating plant construction." Philadelphia Electric then said it was postponing for one or two years its plans to construct two huge nuclear generating stations.

All this represents a drastic change for the utilities, which have had things pretty much their own way for decades—when most people felt the price of electricity was cheap, and the utilities were going all out to urge consumers to use more and more electricity. In turn, the utilities kept putting huge sums of new money into their systems because their rate of return was based on a fixed percentage of their investment. The higher the investment, the higher the profits. And there was no reason to cut costs. Almost everything was justified as an operating expense: legal fees, materials, construction, phone calls, secretaries, and manpower. Regulatory agencies on the state level seemed more like kissing cousins than adversaries—operating in the public interest—of the utilities. The utilities simply asked for increases and mostly got them, until the oil embargo in the fall of 1973. Things haven't been the same since, especially for the nuclear utilities.

Governor Thomas P. Salmon of Vermont, for example, was forced to sign a bill in 1975 stating that both houses of the state legislature would have to give voice approval before the Public Service Commission in Vermont could issue another construction permit for a nuclear-power plant. This stringent antinuclear measure was sponsored by a Ralph Nader-type organization called the Vermont Public Interest Research Group. It passed the state house by 106–39 and the senate by 22–8. A survey showed that over 66 percent of the voters in the state favored the legislation, mainly because they had turned sour on the economies of nuclear power in Vermont.

The state has one 513-megawatt reactor plant—the Vermont Yankee Nuclear Power Corporation Facility, at Vernon, which began generating power in 1972. Vermont Yankee was supposed to cost $80 million; but the completion cost climbed to $220 million. One of two private power companies which together held 55 percent of the stock in Yankee wound up with serious financial problems because of this increase. Power from the plant was supposed to cost 4 mills per kilowatt-hour when plans for the reactor were still on the drawing board in 1966. Instead, average customers found they were paying yearly bills of 18–30 mills per kilowatt-hour for Vermont Yankee power.

Problems at Vermont Yankee also increased the cost of electricity. Fourteen major shutdowns in 18 months were due to accidents, equipment failures, defective parts, and the discovery of dangerous operating conditions. A spokesman for Vermont Yankee, asked to comment on the plant's reliability, was quoted by the *New York Times* as saying, "We're not as bad as some but we're not as good as others."

Nuclear fuel was one of the problems. Radioactive moisture inside the fuel pellets ate through the fuel rods and escaped. The plant had to install a new filter system, at a cost of $10 million, to trap the escaping vapor. Another time the plant was closed for a month and a half because the fuel pellets kept slipping and sliding in the fuel rods, coming together to form "hot spots" that could cause part of the radioactive fuel core to overheat, and perhaps melt. Then the power for a vital safety device was cut off by workmen who were installing a television unit. This caused the reactor to speed up—a very dangerous occurrence under the operating circumstances at the time. Fortunately, automatic emergency controls were able to shut down the reactor before it could get out of hand.

Consumers Power Company in Michigan is another utility with the nuclear money blues. In the early 1970s Consumers Power said it could construct two reactor units at Midland, Michigan, for $349 million —at a cost of $266 per kilowatt—and have them on line by the middle of 1974. Consumers Power began to build. By 1975, to the consternation of the utility's anxious shareholders, Consumers Power said the Midland plant wouldn't start up until 1981 or later, and the reactor units would run to at least $1.4 billion by then—a cost of over $1069 per kilowatt.

The Palisades plant was also draining Consumers. It

cost $95 million more than its original estimate, then ran into a steady stream of problems with its generator and pipes after it began operating in 1971. These, along with vibrating fuel rods, kept Palisades out of action from August 1973 to September 1974 at a cost of some $3 million in repairs—plus an additional expense of around $7 million per month to purchase about 30 percent of the power it was selling from outside sources.

All told, Palisades produced at just a bit more than 1 percent of capacity during 1974. This was also a bad year for Consumers' reputation because the AEC fined the utility for Palisades' excessive radiation leaks back in 1972 and 1973. In turn, Consumers Power sued the Bechtel Corporation of San Francisco and several other defendants for $300 million. In the suit, which is still pending, Consumers charges Bechtel—which is also building the Midland plant—with negligence in the design and construction of Palisades. Lawyer Mike Cherry kept pressuring the Michigan Public Service Commission to reexamine its decision to let Consumers generate nuclear power on the basis of its record at Palisades and its expensive, time-consuming difficulties at Midland.

Some nuclear utilities and manufacturers with financial problems have asked the federal government to either underwrite their loans or simply pick up the tab. Consumers' chairman did the former; he asked the federal government, but without much luck, to purchase enough of the company's stock to finance $200 million in capital expansion. Westinghouse went the other route in asking Uncle Sam to buy Westinghouse's floating nuclear-power plants.

This leads us to Jacksonville, Florida, the scene of a multibillion-dollar story of nuclear tug-of-war with enough dizzying twists and turns to intoxicate a mystery freak, along with confirming some Watergate-inspired suspicions about corporate dealings in the land of high stakes. It all began in the spring of 1972. Public Service Electric and Gas Company of New Jersey, interested in reactor plants but short of land on which to put them, approached Westinghouse with the idea of building floating nuclear-power stations and anchoring them offshore to supply the mainland with electricity.

Westinghouse, intrigued by the possibilities, teamed up with giant Tenneco to form an outfit called Offshore Power Systems. Westinghouse would build the 1100-megawatt reactor plants; a Tenneco subsidiary, Newport News Shipbuilding and Drydock Company, the nation's largest shipyard, would design and build the hulls for the plants, which were going to be priced at $500 million and up. The idea was to construct them on land, tow them to sea, surround them by an immense breakwater, and anchor them.

On paper it must have seemed very logical to the developers. The floating nuclear plants would be relatively inexpensive, as reactors go; they would, for instance, avoid the cost of buying land; their proponents claimed the environmental impact would be negligible. The ecologists, however, thought it would be a nightmare—nothing less than total madness, technology run amuck, a scheme that ran the risk of polluting the seven seas with radiation for centuries in the event of a major accident at one of the floating nuclear plants. A number of communities were trying to attract OPS; but after considerable wooing on the part of Jacksonville's business elite, Offshore Power Systems decided to build a $300 million construction facility at Blount Island, midstream in the St. John's River, 10 miles from Jacksonville and an equal distance from the ocean.

Jacksonville thought OPS would create more than 10,000 new jobs with an annual payroll of $160 million, raise real estate values, and give the city $8 million in new tax revenue annually. The city establishment also thought the community would bask in countless other financial blessings as the brilliant center of a sparkling new growth industry—which was supposed to send four floating nuclear plants out to sea each year to customers who would be eager to buy them. But things didn't turn out that way. Jacksonville—for a long time—may have been enthusiastic about the floating plants, but utilities around the country were not.

Jacksonville handed OPS more than 800 acres of prime local real estate; began to build an $11 million vocational education center to train workers for the OPS plant; and offered to sell $180 million in municipal bonds to supply OPS with on-hand cash, and build a $137 million bridge to Blount Island. In addition, the municipally owned Jacksonville Electric Authority offered to buy two floating nuclear plants from OPS for $2.2 billion, and signed a $4 million enriched uranium fuel contract with the AEC for the plants JEA was going to buy for Jacksonville, plunking down $1.4 million with the AEC to seal the deal.

In retrospect, the affair between Jacksonville and OPS seems like a cozy interlocking match between the company and a number of local municipal authorities whose members included business and community leaders who had ties with OPS, and stood to profit, either directly or indirectly, from the now fading dream to build the floating reactor plants. One thing they didn't count on was tangling with an extremely tough, tenacious gentleman by the name of Joseph H. Cury, a grocer by trade, who found a consuming passion as the head of a local group called POWER—an acronym for People Outraged with Electric Rates. Joe Cury, who wound up in a no-holds-barred battle with OPS and the business-oriented power structure in Jacksonville, claims that he and members of his family were threatened with bodily harm and that attempts were made to smear him, blackmail him, and drive him out of business and out of Jacksonville. The FBI was called in on the case, which is packed with intrigue, melodrama, threats, and financial skulduggery. One of his finest moments, according to Cury, was a 12-hour marathon debate he led at a city council meeting in Jacksonville, during the course of which he and others were finally able to talk the city out of buying the $2 billion floating nuclear-power plants from OPS.

Chunky, dark-haired Joe Cury got involved in his tussle with Jackonville's business elite and OPS after 20 years as a popular, prospering grocer and a complacent electric ratepayer back in the days when Jacksonville was burning inexpensive imported oil to produce the cheapest municipal power in the state. Then his electric bill bounced up—way up—and Cury wanted to know why. In December 1973 Cury paid $45 a month for electricity at home and $700 for power at his supermarket. One month later his home bill was $95 and his business bill was $1500 a month.

JEA had its own story. In 1973 JEA was burning 10 million gallons of oil at $2.69 a barrel. Electric bills were just about the same in Jacksonville as they had been in 1942. When JEA's contract for oil expired in September 1973 the authority reportedly sent out more than 40 bids for a new contract. No replies. Desperate, JEA entered into an agreement to buy Venezuelan oil at a starting price of $5.25 a barrel, with a clause that permitted the price to rise—which it did, to $12 a barrel by January 1974, when Joe Cury and other ratepayers in Jacksonville discovered their electric bills were doubling.

Cury claims the men who served on JEA gave him the brush-off when he pressed for detailed explanations about the rate hikes. He began to investigate, and he made loud, aggressive charges against JEA, accusing its members of irregularities and worse in their purchase of fuel oil in particular and in their operation in general. Largely at Cury's instigation, a federal grand jury began to investigate the price of oil and JEA. By early 1976, the grand jury was still investigating but had handed down no indictments. Meanwhile, JEA thought it could solve the problem of its skyrocketing fuel costs right in its own backyard—by purchasing floating nuclear-power plants. JEA proposed to finance the $2.2 billion deal by floating that amount in municipal bonds. A few local critics said it would be the largest tax-exempt debt ever shouldered to finance any project in the United States—which could bankrupt Jacksonville if anything went wrong.

Cury, highly suspicious by then of anything connected with JEA, began to investigate the pros and cons of nuclear power, and concluded that the floating nuclear plants were far too expensive and far too dangerous. By mid-1974 Cury had a full-blown campaign going to squelch the proposed deal between JEA and OPS, which soon developed into a contest between the local establishment (for) and a growing segment of the nervous public (opposed). Cury went to Washington, to the state capital, attacking everything and anything connected with OPS, the Jacksonville municipal authorities, and the city's business establishment.

"I got so involved," Cury said, "that I couldn't back down." Cury alleges that Westinghouse first tried to wine and dine him into complacency but soon switched its tactics as the matter "developed into an all-out war between Westinghouse and myself."

One example was his candidacy for city council, a primary race he lost by a narrow margin in a runoff election held in 1975. "Usually a candidate will spend about $500 campaigning for this office, but they spent $60,000 against me!" Cury alleged. Money wasn't the only weapon Cury faced. He is half-Lebanese and half-Irish by descent, so word was spread alleging that he was an Arab fanatic and anti-Jewish, and next that he had a police record—for a misdemeanor conviction back in his hometown of Allentown, Pennsylvania, more than 20 years before. Cury charged that a Westinghouse official tried to make illegal use of a state police agency to obtain information that he had pled guilty to an old conspiracy-to-robbery charge in Allentown for which Cury was fined $100.

Cury says the more he was pushed, the louder he yelled. He denounced everything connected with the floating power plants. "Westinghouse knows I'm not going to back down," he told me. "They know I'm not going to let them build the plants." Cury's aggressive opposition helped torpedo OPS's sale of the floating reactors to Jacksonville. This was only part of the bad news for OPS in the summer and fall of 1974, which at one time had eight orders for floating plants. A utility in Louisiana dropped its plans to buy two. Next Jacksonville pulled back. OPS was down to four orders. Then New Jersey Public Service and Gas, which was going to take four of the eight floaters, said it would have to delay its orders at least five years. Finally, in early 1975, Tenneco withdrew from OPS, which became a subsidiary of Westinghouse alone.

The only customer in sight—maybe—for OPS was the New Jersey utility. Jacksonville, tentatively stuck with a lot of big bills and projects under way, turned sour on the whole venture. OPS reportedly had put at least $50 million into its half-completed Blount Island construction yard, expected to cost over $150 million before it was finished. So OPS has gone looking for "Uncle Sugar"—as *St. Petersburg Times* columnist Dudley Clendinen put it—to buy four to eight floating plants and then sell or lease them out to electric-power utilities "so they will be ready when the energy crunch comes," according to Westinghouse Board Chairman Robert E. Kirby in April 1975.

But the electric companies have been having their own financial problems. This is the industry that Federal Energy Administration official Eric R. Zassner in 1975 said was "near financial collapse," a condition that can be attributed in part, at least, to the economics of nuclear fission plants—with critics like David Comey and Jim Harding sniping so effectively at nuclear plants on the economic issue.

It was Jim Harding, of Friends of the Earth who in October 1975 was credited by the Investors' Responsibility Research Center's *News for Investors* with providing arguments that led to the Sacramento Muncipal Utility District's decision to table its approval of a half interest in a nuclear plant and opt instead for conventional combustion turbines, and the purchase of hydro and geothermal power. Harding, who testified in April 1975 before the directors of the Sacramento utility on the proposed Rancho Seco 2 nuclear generating station 25 miles southwest of Sacramento, pointed out many of the economic liabilities for nuclear plants that were echoed in January 1976 by William C. Walbridge, general manager of the utility.

The liabilities cited by Harding, later expressed by Walbridge, were becoming hauntingly familiar to the nuclear establishment: increasing capital costs (from $427 million for a half interest in Rancho Seco 2 to $660 million in two years' time, according to Walbridge); the increasing price of uranium; the disappointing operating (capacity) experience of nuclear plants; uncertainties related to fuel enrichment; the lack of reprocessing plants; and the absence of a "definite program for the ultimate disposal of high-level radioactive wastes," in Walbridge's words.

In his 1975 testimony Harding, who argued that the "alternative technologies of coal and geothermal power could easily be lower in cost" for the Sacramento utility, urged the company to "study more carefully the prospects in energy conservation that could make more efficient use of current . . . generating capacity. Staggered work hours at commercial and industrial facilities and insulation programs for all Sacramento area buildings could cut sharply into current and projected demand. The cost of saving 1 kilowatt is almost always less than that of adding 1 kilowatt of supply," Harding said.

Walbridge was talking along the same line ten months later. "It is obvious that this country uses far more electricity per capita than most other major industrial countries, that much of this electricity is wasted, and that programs to reduce this waste should become a major part of this country's efforts to alleviate the energy crisis. . . . What we intend to do is to bring to the attention of the appropriate federal and state authorities, as forcefully and frequently as seems practical, the need to place mandatory conservation programs into effect."

And Comey and Harding and Ralph Nader—who by November 1975 was saying that a "de facto moratorium" had already been imposed on the construction of nuclear plants because of crippling economic factors —were just a few of the many critics who were convinced that they had enough ammunition almost to defeat nuclear power on the economic issue alone.

VIII
The Citizens' Battle

Some of the most bitter, instructive battles in the controversy over nuclear power have been fought on the local level. One of them was sparked by a bolt of lightning. It must have been very frightening. Emergency warning systems began to flash. Sirens started to wail. Men who were working at the nuclear construction site froze in their tracks. But when they heard a foreman yell, "Get the hell out of here!" they broke and ran, scattering for their trucks and cars. Moments later, vehicles were racing away long back country roads and over the one-lane bridges that wound away from the west banks of the Susquehanna River.

This was in July 1971 at about 7:00 p.m. Pennsylvania farmers and their families were done with their dinners and chores and were sitting on their front porches. They heard the siren from Philadelphia Electric Company's Peach Bottom nuclear fission power complex in southern York County, and watched the cars and trucks careening by.

They were really worried, though the utility had assured them from the beginning that nuclear power was as safe as apple pie—had wooed them, in fact, at the Peach Bottom Visitor Information Center with generous servings of homemade apple pie, ice cream, and soft, soothing talk. The next day, the Pennsylvania Dutch learned that the lightning had struck a power line, which triggered an emergency alarm system. But people living near the nuclear-power complex were still uneasy, and not only because of their fears the night before that one of the nuclear plants had "blown up" as they put it, but also because they had not been notified by Philadelphia Electric when the site was abandoned.

"The only thing they were thinking about was saving their own necks," I was later told by one disgruntled farmer, who said he was also concerned about the possible effects that routine discharges of low-level radiation from the Peach Bottom nuclear reactor complex might have on his wife and three growing daughters, his crops, livestock, and orchards. When asked if he ever worried about a major accident at Peach Bottom, the farmer threw up his arms in frustration. "They say it can never happen," he said.

A year later, toward the end of July, after a thunderstorm on a hot sultry night, people living in southern Lancaster County, on the opposite side of the Susquehanna River from Peach Bottom, were startled by the sound of a loud "whoosh," which persisted for an hour or more. A woman who lived in the area told the press she thought it was "the world's largest steam kettle letting off steam." Then the lights went out. People made their way to phones. Calls, demanding explanations, began to flood the utility switchboard at Peach Bottom. Philadelphia Electric employees insisted, honestly as it turned out, that no one on their side of the river—dammed to form a pond here that is 1½ miles wide—had heard the noise as it was being described to them.

A few days later the plant handed out a press release. It said lightning had struck again, this time hitting a transformer and knocking out a generator, which permitted the stream of steam to escape with a giant wheeze.

People in the area were beginning to get more and more queasy about the two giant reactor units that were going up at Peach Bottom next to the small 40-megawatt experimental high-temperature gas-cooled nuclear plant that had been generating electricity since 1967. And just 25 miles upstream, in the northeast tip of York County—just south of and across the river from Harrisburg, the state capital of Pennsylvania, at a place called Three Mile Island, there was Metropolitan Edison Company, also busy building two large nuclear reactor units.

Philadelphia Electric announced it was going to build another set of twin reactor units—Fulton 1 and 2—in Lancaster County, right across the river from Peach Bottom. These would be 1140-megawatt gas-cooled, high-temperature units, modeled after the experimental plant at Peach Bottom, only a thousand times as large in generating capacity. People in the southeast corner of Pennsylvania were further disgruntled to discover that Philadelphia Electric had been talking with authorities in Maryland about the possibility of building still another nuclear plant just south of the Pennsylvania-Maryland line. All this, if it came to pass, would plunk at least eight nuclear fission reactor units along a 35-mile stretch of the Susquehanna—five of them within 2 miles of one another, which would make the York-Lancaster area the largest concentrated source of nuclear power in the world.

Some people were against nuclear power, no matter where. Others were opposed to nuclear plants in their own backyards. More were willing to accept one nuclear reactor unit—even two or three, but not five or more within close proximity, not if they were going to make York-Lancaster the nuclear capital of the world, as the local critics warned they would.

Two groups came out of this local trepidation to fight the reactors—the York Committee for a Safe Environment on the west side of the Susquehanna and the Save Southern Lancaster County Environmental Conservation Fund on the east. Their stories each illuminate a particular time, a mood, a setting in the controversy over nuclear power.

In 1972, when the grass-roots movement against nuclear power in York and Lancaster began to develop, the battle seemed like a contest between David and Goliath—which in many respects it was, both on the local and national level. One side of the barricades was manned by fat, well-paid troops employed by the likes of Westinghouse, General Electric, General Atomic, Exxon, Kerr-McGee, Combustion Engineering, Babcock and Wilcox, nuclear utilities like Philadelphia Electric, and the AEC—resoundingly supported by the powerful U.S. Joint Committee on Atomic Energy.

The multibillion-dollar nuclear power establishment really seemed to be riding high in 1972. Every national newsmagazine, or so it seemed, was running ads promoting nuclear power; some cost as much as $40,000. They were smooth and bland and full of assurances that nuclear fission was bound to be our principal source of power in the future.

The antinuclear movement was an informal coalition of scientists, environmentalists, and other critics scattered around the country, without any money to speak of, but with a good deal of distinguished expertise; these informed critics kept repeating that nuclear fission power was so loaded with problems, on such a horrendous scale, that the spread of nuclear plants could doom civilization as we know it; and not only for today and tomorrow, but as far into the foreseeable future as the mind could imagine. Among the leading antinuclear groups at this time were the Cambridge-based Union of Concerned Scientists, with a hundred members, most of them on the faculties of Boston-area schools; Friends of the Earth, with 20,000 members in 50 states; and the Committee for Nuclear Responsibility, on the West Coast, with about 1000 supporters.

All this activity was beginning to seed. By the end of 1972 the tide was beginning to turn, if only slightly, against the nuclear establishment. Consumer advocate Ralph Nader, for one, impressed by the work and arguments of the Union of Concerned Scientists, was on the Dick Cavett TV show in December 1972, saying, "There is no question now that we have to have a moratorium on the construction of nuclear-power plants in this country." The *New York Times* also gave the critics a boost by editorially declaring on January 31,

1973, "Once so promising in the first enthusiasm of the atomic era, nuclear power is becoming something of a monster, with dangers to people and the environment so awesome as to raise serious doubts that this is indeed the best energy source of the future."

Encouraged, the York nuclear critics hired local attorney Raymond Hovis to represent their case before the AEC in an effort to stop Philadelphia Electric from obtaining an operating license from the agency for the two new giant reactor units at Peach Bottom. Ray Hovis, then 40 years old, was an ex-member of the state house, the nephew of former Pennsylvania Governor George Leader, and a very thorough, careful, conscientious lawyer who had a local reputation for being open-minded, honest, and fair. Hovis started out with a healthy suspicion of Philadelphia Electric, based on his experiences as a state representative with the heavy-handed influence the utilities have traditionally wielded in Pennsylvania politics; but he had no decided opinions about nuclear power in general and the AEC in particular.

"The AEC handed me this," Hovis later told me, pointing to a row of thick volumes in his law offices in downtown York, a city of 50,000 people with twice that many in the suburban belt that surrounds the county seat. "I had to go through 6 to 8 feet of this reading material just to get started on Peach Bottom." By the fall of 1973 Hovis was tapping his desk with annoyance. "I find the whole thing very educational," he said with a wry smile. "I'm not totally opposed to nuclear power; however, I'm really afraid that we're going to become a receptacle here of nuclear-power plants. As far as Peach Bottom goes," Hovis said with a sigh, "the only thing we can fight for when it comes to Philadelphia Electric and the AEC is to try to force them to make the plants as safe as they can be under the circumstances."

What Hovis was learning, as a result of the AEC hearings on Peach Bottom, was common to most intervenors, as the people who oppose nuclear facilities at hearings are called—that it is very difficult for any individual or citizens' group to stop the construction and operation of a nuclear-power plant. The nuclear establishment, on its part, learned that it was self-defeating to try to build a reactor plant right next to the San Andreas fault of San Francisco earthquake fame, or inside a city (New York) or immediately outside one (Philadelphia and Newark). But there were only 8000 people living within a 5-mile radius of Peach Bottom, out of sight, and in the beginning at least, out of mind for the city of York, 35 rural miles away; and also for the Quaker City 65 miles to the north, where most of Philadelphia Electric Company's million-plus customers lived. Besides, Philadelphia Electric (and the AEC) said the plants would be safe and a boon to the York economy.

Local Congressman George A. Goodling, a conservative Republican and a popular figure, accepted the nuclear plant by saying, "If it's good enough for the AEC, it's good enough for me." The York Chamber of Commerce and Industry, along with the local building trade unions, blessed the new $900 million units because of the construction jobs they would provide and the amount of money they were expected to pump into the local economy.

Ray Hovis raised a number of objections to the Peach Bottom reactors before the AEC Safety and Licensing Board hearing held in York in the summer of 1973, among them the following: nuclear plants were proliferating in the York area at an alarming rate; the AEC was refusing to consider the cumulative effects of low-level radiation that would be released by the three reactor units at Peach Bottom; the agency had ignored the risks involved in transporting radioactive fuel to and from the reactor complex, and had failed to compel Philadelphia Electric to produce a realistic, workable emergency evacuation plan in case of a nuclear accident. Hovis said the licensing board dismissed all these objections as "irrelevant" in 1973, when it granted a conditional operating license (pending the long-term installation of closed circuit cooling towers at Peach Bottom) to Philadelphia Electric. Hovis appealed the hearing board decision to the AEC appeals board. It eventually ruled against the intervenors, which by this time included the antinuclear group in neighboring Lancaster County.

Hovis took the case to the U.S. District Court of Appeals in Washington, D.C. It was argued in November 1975. The court handed down its decision the next month, which said, in part, "We find no flaws in most aspects of the commission proceedings." The court did feel, however, that the discharge of radioactive materials from the reactors, in particular radioactive iodine 131, might not be consistent with the commission's standards of as "low as practicable." The court then said the reactors could continue to operate but ordered the federal agency to "determine in appropriate proceeding whether to modify the operating license for the Peach Bottom reactors to require additional emission control equipment." The court also touched on another point raised by the intervenors—their request for financial and technical assistance from the commission, which the agency initially rejected, but then said it was reconsidering. The court declined to rule on this issue, or "express any views regarding financial assistance during the remand proceedings," then added, "We note, however, that it would be unrealistic to expect public interest litigants to underwrite the expense of mounting the kind of preparation and presentation of evidence that is ordinarily required in this type of case."

Chemist Chauncey Kepford of York was one of the most active members of the local antinuclear group. A passionate, indignant man, who wound up waging a personal vendetta against the nuclear establishment, Kepford had this to say in 1973 about the AEC hearing process, based on his frustrating experiences in York and elsewhere: "We're strapped to the AEC's quasi-legal format, which means we can't stop these plants because the AEC in general is in the crazy position of being advocates of nuclear power, who assume it will be the normal means of power generation in the future. At most, all these guys are interested in is getting what they call 'proper technical regulation.' "

Dr. John Gofman of the Committee for Nuclear Responsibility elaborated on this theme when he said, "Nothing has suited the promotional nuclear-power interests better than keeping alive the misconception that a decision pro or con nuclear fission power rests upon esoteric technical arguments [at] so-called 'public hearings' [where] concerned citizens have been led, like lambs to the slaughter, into the promoters' arena to contest a variety of valves, filters, cooling towers, and miscellaneous other items of hardware in specific nuclear plants.

"But this is not where the problem lies . . . what is really at issue is a moral question—the right of one generation of humans to take upon itself the arrogance of possibly compromising the earth as a habitable place for this and essentially all future generations . . . visiting cancer on this and a thousand generations to come [and] the prospect of genetic deterioration of humans that will ensure an increase in most of the common causes of death in future generations."

In 1974, in a letter to me, Dr. Gofman added this footnote: "The nuclear technologists (and I've been one in the past) live in the dream world that they can calculate the incalculable. The only ingredient missing in their calculations is a small sprinkling of common sense."

Raymond Powell represented the opposite point of view when I met him in 1974 at the AEC's headquarters in Bethesda, Maryland. Powell, a nuclear core physicist, was the AEC licensing project manager for the Peach Bottom reactors, and in 1975 was still serving in that capacity for the NRC. "This is the most regulated industry in the nation," Powell said in 1974. "If everyone met the kind of standards we impose, you'd hear no more about consumer complaints. We're very strict when it comes to basic specifications," Powell said. "You have to be," he insisted, "when you're talking about public safety."

But critics like Hovis complained in 1974 that "the whole trouble with the AEC—and the utilities—is they say, 'We'll meet whatever regulations exist.' But the AEC establishes these regulations. These people are judge, jury, and advocate all rolled into one. And it's like wrestling with an octopus waving arms of rules and regulations all around and in and out. When we tried to talk about low-level radiation discharges we were told that the emissions would fall below the federal threshold for each reactor. But this criterion doesn't take into account the cumulative total amount of radiation coming out of Peach Bottom. It also ducks the validity of these federal standards.

"And whenever we tried to raise the issue of future problems," Hovis continued, "they assured us that when the time comes, they'll be taken care of. But what do you do, for instance, with a 'hot' nuclear plant? They're only licensed for 40 years. After that we're told they get too 'leaky' to run. What do you really do to decommission a hot nuclear plant? And there's going to be hundreds of them around if this keeps up," Hovis said, showing me a thick AEC manual on Peach Bottom that contained an estimated figure of $100 million to completely decontaminate a 1100-megawatt nuclear plant.

This led to Hovis's complaint that too many people employed by the nuclear establishment have lost a good deal of the perspective he feels they should have. "These people who work for the AEC and the utilities are the victims of their backgrounds," Hovis said. "They've been working with nuclear power for so long, many of them, and have so much of their lives invested in it, that they have been sold on their own promotion. Nuclear power has only been around for 30 years or so and these people feel they know a hell of a lot more about it than anyone else."

Raymond Powell says he has spent his whole professional career, which began in the early 1950s, working in the field of nuclear energy—"common," Powell told me, "for 75 percent of the senior AEC people." Powell came to the AEC from AMF Corporation's atomic division—located, by coincidence, in York. Like other people whom I've met who work for the nuclear establishment, Powell seemed to be an exceedingly able white-collar-and-tie man who was dedicated, conscientious, hard-working, serious, deliberate, and self-contained. All of this was underscored by a deep faith in nuclear power that seemed closer to certainty than mere confidence; this too is characteristic of many people in the atomic field.

Chauncey Kepford was even more certain than Powell—in Kepford's case, that nuclear power was too dangerous. A few years before I met him in 1973, when he was 37 years old, Kepford was teaching chemistry at the Pennsylvania State University extension campus in York, living in a split-level house in the suburbs and dressed like his neighbors. Before this he had been a radiation research chemist for a New England laboratory. One day Kepford came across a story in a York newspaper reporting that Metropolitan Edison Company had applied for a permit to dump 50,000 picocuries of tritium—a heavy radioactive isotope of hydrogen—per liter of water into the Susquehanna from the Three Mile Island nuclear plants. Curious, and thinking this was excessive, he publicly questioned the proposed discharge. The amount was lowered from 50,000 to 500 picocuries. Kepford decided to take a good look at nuclear fission power.

"I started off by reading popular paperbacks on nuclear power," Kepford said. "But I was still kind of casual about it. Being a reference freak, I started checking up on the AEC and other government stuff. And what I found just blew my mind." Kepford became obsessed about the potential dangers of nuclear power. He began to speak up, and loudly. He became controversial and lost his teaching job. Claiming that the nuclear establishment had applied pressure on Penn State to have him dismissed, Kepford hired a lawyer; an out-of-court settlement was reached giving him "enough money to keep going," he said, and the freedom to devote all of his time to fighting nuclear power. Then his wife left him, taking their three young children with her. "My wife was against nuclear power too, but not that committed," he said, waving a hand toward his shelves of books on nuclear power. "I have to keep fighting it," Kepford said. "It has to do with self-respect."

He moved from the suburbs into York's inner-city area, into a row house by the railroad tracks, let his hair grow long, grew a beard, got a van, changed into jeans, and feeling free and obsessed, went far and wide to fight against, testify against the nuclear establishment, which Kepford urgently believes must be defeated before some catastrophic accident occurs.

The lack of money finally became quite a problem for the antinuclear group in York. Hovis, hoping to be paid for his legal work on Peach Bottom, has only received a token amount in fees for the hundreds of hours he has devoted to the case. In December 1974 Hovis said the York group owed him around $21,000. And he estimated it would really take a minimum of $50,000 to put a good legal case together against the nuclear establishment. "It's almost impossible for the public to get involved without some kind of economic assistance," he said.

Charles Bacas, another antinuclear critic, who lives in York with his wife and two sons, looked at the AEC hearing process with a mixture of disgust and concern. Bacas, a former newspaper reporter who went to work as the executive assistant to the secretary of the State Department of Community Affairs in Harrisburg, contends these government-run nuclear proceedings have worked to deny citizens their democratic right to choose. Thoughtful, well-read, and intense, Bacas wrote this down for me: "The process of assessing technologies has got to become part of the democratic process, or in the coming decades our liberties are going to decline as drastically as technological complexity is expected to rise. At the present time, our votes at the polls and our dollar votes in the marketplace have a minimal effect on what are essentially elitist 'planning' decisions, whether by government or business. A number of risk-taking technological programs have been implemented, with the result that the democratic process has been particularly ill served by the governmental and corporate proponents of these technologies, because potential dangers have been cloaked behind the man-in-the-white-coat certainties of deterministic science.

"This science no longer exists, and its certainties are no more available to us than the once seemingly limitless American frontier. Yet scientists and corporate and governmental executives continue to make particular decisions affecting the well-being of us all without even having the vaguest general consent from the public to do so."

On the east side of the Susquehanna, members of the Save Southern Lancaster County Environmental Conservation Fund were in better shape from the time they began to protest than the critics in York; and the Lancaster antinuclear movement continued to improve the longer their fight went on. One important factor in their favor was that they began to oppose the Fulton reactor units long before Philadelphia Electric was in a position to apply for an AEC construction permit to build the 1140-megawatt high-temperature gas-cooled reactors, initially estimated to come on line in 1981 and 1983. Philadelphia Electric was at a disadvantage because it didn't own 600 acres of land at Fulton as the law required for a nuclear plant: a number of the local property owners did not want to sell to Philadelphia Electric. But the utility, in 1973, was determined to build the plant at Fulton, where it already had rights of way for its high-tension power lines; moreover, the Susquehanna—the largest river east of the Mississippi—seemed to have plenty of water to disperse waste heat from the reactors. Peach Bottom Units 2 and 3, for example, were drawing and discharging a billion gallons of cooling water a day from the river.

Philadelphia Electric soon learned that things were going to be much more difficult at Fulton than they had been at Peach Bottom. The Lancaster intervenors, drawing upon their observations of York, were much stronger and better organized. They had over 100 member—farmers, housewives, professional men and women; and they had enough money to put up an effective fight against the utility. They also had members of the faculty from Lancaster's small but prestigious Franklin and Marshall College on their side—geologists, physicists, chemists, and engineers who were able to supply the Lancaster critics with valuable information and testimonies.

Public opinion was on their side too: "We collected petitions at the local county fairs during the summers and found sentiment running 10–1 against the reactors here," I was told by George L. Boomsma, president of the Lancaster group. Boomsma was the driving force behind the opposition to the Fulton reactors. He turned himself into something of an expert on nuclear power in particular and energy in general and kept passing out provocative information in a highly informed monthly newsletter. Boomsma is a salesman who used to live around New York City. In the late 1960s, in search of clean air and fed up with urban congestion, Boomsma and his family found an old stone farmhouse to renovate in the wooded hills along the east bank of the Susquehanna. Apprehensive about the Peach Bottom reactor units across the river, Boomsma and his wife were much opposed to the construction of a nuclear plant on the very fringes of their land.

They tried to learn as much as they could about nuclear power in a hurry, and were soon convinced that the Fulton nuclear plant would endanger the safety and health of their growing children and everyone and everything else in the area.

George Boomsma, at first impression, seems like an ordinary smooth-faced man in his forties, with short hair, a friendly smile, and an easygoing manner. Most people meeting him for the first time would probably use the word "reasonable" to describe Boomsma. He is a good listener, he's deliberate and careful. And after a short while they would also discover that he is deceptively bright and shrewd; that his warm, folksy touch tends to disguise a rather unbelievable determination, which he focused against nuclear power.

By mid-1973, people on both sides of the river, and downstream too, were in a mood to take a much harder look at the Fulton plant than they had at Peach Bottom, where a farmer's wife near the York plant sighed as she said, in retrospect, "It was too big for us to fight here. When we made a ripple, the Philadelphia Electric public relations people just came in and snowed us. So 95 percent of the people around here thought atomic power was OK. Their public relations people did a wonderful job. We all got invited to the plant. They fed us homemade pie and ice cream—and spent all their time telling us there was nothing to worry about." The Peach Bottom Township Board of Supervisors in York County joined with Fulton Township in filing formal protests against the proposed reactor complex at Fulton. Congressman William Goodling in York, elected after his father retired, was of a different frame of mind than George Goodling. "I have a lot of concern and reservations—about nuclear waste and safety," Bill Goodling told me on Capitol Hill in May 1975. "And there's too many reactors around York County. It's time to put 'em someplace else." Congressman Edwin D. Eshleman on the Lancaster side felt the same way. "And hardly anyone in Lancaster publicly came out in favor of the plant," Boomsma said.

The city of Baltimore, which draws a large percentage of drinking water for its 2 million people from the Susquehanna below Peach Bottom, and could be affected by a major nuclear accident upstream, was also expressing increasing concern. So was an environmental group called the Chesapeake Bay Foundation. In order to protect the ecology of the Susquehanna, to prevent hot water from Peach Bottom from overheating water in the river, which could kill fish and other forms of marine life, the AEC was asking the Peach Bottom nuclear complex to install closed-cycle cooling towers before 1977 at a cost of well over $100 million. And the Fulton reactor units were also required to incorporate a closed cooling system into their design. These cooling towers evaporate tremendous quantities of water, which in turn could upset delicate ecological balances even in a body of water as large as the Chesapeake Bay, which draws 47 percent of its freshwater input from the Susquehanna.

The Fulton reactors would evaporate some 28 million gallons of water a day. This loss and the evaporation from the other nuclear plants on the river would approximate one-third of the Susquehanna's low-water flow. A sharp reduction in the amount of fresh water emptying into the Chesapeake Bay could ruin commercial fishing. Oysters, for example, thrive on a blend of fresh water and salt.

Philadelphia Electric seemed somewhat carefree about this concern over the future of the Chesapeake Bay, contending that any problem that arose could be handled without much trouble. The nuclear critics were afraid it might be too late to repair the damage if the utility was wrong. This difference of opinion is typical of the nuclear controversy, in that the nuclear establishment has displayed almost boundless confidence in its ability to solve all problems and contain all hazards should they arise.

People in the York-Lancaster area were also concerned about the type of reactors slated for Fulton, and were disturbed by the difficulties and shutdowns Metropolitan Edison and Philadelphia Electric were having with their reactor units along the Susquehanna.

The high-temperature gas-cooled reactors slated for Fulton would use highly enriched uranium fuel—weapons-grade material from which atom bombs are built—and at 1140 megawatts would be vastly larger than any plant of its type—especially the 40-megawatt prototype at Peach Bottom which Philadelphia Electric shut down in October 1974 after saying its commercial output was too small to be an asset. The concern here was this: the model has held up at 40, but will it at 1000? And by 1974, and into 1975, every news outlet in the area seemed to have at least a report a week or so on another nuclear mishap or danger. There were valve failures at Peach Bottom. Metropolitan Edison made headlines for lax security at Three Mile Island and excessive discharges of radioactive gas. Philadelphia Electric's Peach Bottom reactors were among 21 in the nation that were ordered to shut down in January 1975 to look for possible cracks that had been developing in the cooling system of boiling-water reactors. But the most serious problem that Philadelphia Electric was running into over Fulton was money, because the estimated cost of the plant had risen from $1.4 billion to $2.5 billion; operating dates had been pushed back by the utility to 1984 and 1986, and even this timetable was subject to doubt.

By the summer of 1975 the NRC had reviewed Philadelphia Electric's application to build the reactors at Fulton and had issued a safety evaluation report asking for changes in its design and more additions, which further drove up the costs for the plant. There was a growing doubt about the need for it. Then on October 23, 1975, the only supplier of high-temperature gas-cooled reactors in the country, which had built the 40-megawatt Peach Bottom 1 unit and had been dealing with Philadelphia Electric at Fulton, announced that it was leaving the commercial reactor business "for the time being." This seemed to end the Fulton nuclear project, at least for the moment.

The firm, General Atomic Company of San Diego, California, a wholly owned joint subsidiary of Gulf Oil Company and Royal Danish Shell, had reportedly lost about $500 million in the reactor business by the time it announced its decision to retire.

"In the long run," Boomsma said, "economics and what is practical will determine the future of nuclear power. It's not only dangerous but it's unreliable and impractical; and it's going to prove way too expensive in the long run for the public to put up with."

Peter Anderson, a young consumer-minded conservationist and lobbyist for a group called Wisconsin's Environmental Decade, was saying much the same thing around the same time. "When I was a kid," Peter told me in Madison, the state capital, "people looked upon technology and 'progress' as a god; but today, people are losing their religious faith in technology, in nuclear power. And when people begin to learn about the enormity of the problems connected with nuclear power, their eyes go wide open. So there's an audience in Wisconsin today and elsewhere that's really open and receptive to the whole issue of nuclear power and energy. And the more they actually learn about nuclear power, the more turned off they get."

Attorney Mike Cherry from Chicago was coming to the same conclusion about the nuclear climate in Wisconsin, or perhaps it would be more apt to say that many people of Wisconsin were being turned around on the nuclear-power controversy by Mike Cherry.

The nuclear battle in Wisconsin revolved around a 1974 proposal by four power companies in the state, led by Wisconsin Electric Power Company, the largest in Wisconsin, to build a $1.2 billion nuclear-power plant, featuring two twin 900-megawatt Westinghouse reactor units at Lake Koshkonong, a large, shallow, mud-bottom lake near the small agricultural town of Fort Atkinson, about 30 miles southeast of Madison. The four utilities, which together supplied Wisconsin with 77 percent of its electricity, had long been accustomed to having their own way. William F. Eich, a young conservation-minded attorney who had been named chairman of the State Public Service Committee by Wisconsin Governor Patrick J. Lucey, put it this way in the fall of 1974: "In the past, the utilities liked to regard the PSC as their partner in assuring there was enough energy to meet the needs of the people. Now we're showing a brand-new skepticism toward them. And they may feel that the ground, which has always been so firm under their feet, is a little shaky. And when Mike Cherry gets involved," Eich told me, "anything can happen. It's a whole new ball game."

But in the spring of 1974 before the hearings began, the Wisconsin Electric and its partners were confidently predicting that construction at Lake Koshkonong would begin in September 1975, that the plants would be generating by 1981, and that they were desperately needed because there was going to be a 59 percent increase in electrical consumption in the area the utilities served by 1981. Although their estimates were way off on all counts, the Wisconsin utility consortium had every reason to feel complacent—until it ran head on into Cherry.

There were four nuclear-power plants in Wisconsin. The first one, built by an electric co-op, was a small experimental unit. The next three were large commercial reactor plants owned by the big utilities in Wisconsin. These three plants were approved by the PSC within a matter of months. By 1974 nuclear power was supplying Wisconsin with about 35 percent of its electricity. But the times were changing, and fast, even though the utilities in Wisconsin didn't seem aware of it at first. Farmers were becoming increasingly unhappy with the huge high-tension power towers that the utilities kept stringing across valuable farmland in southern Wisconsin; and their protests were beginning to prod the long-dormant PSC with reminders that the regulatory agency had been established in 1931 to act in the public, not the private, interest.

And the electric utilities had a paternal, far from candid attitude that was beginning to generate irritable post-Watergate sentiments in large segments of the public. People began to complain that the PSC and the utilities were in tandem to boost their electric bills (one of the four power companies involved in Koshkonong had asked for and received two rate hikes in 1974 alone) without due regard for the public interest. This fed a latent suspicion that Wisconsin Electric and its partners wanted to build Koshkonong to increase their capital investment which in turn could be used as a basis to ask for more rate increases.

Also, there were only 4 million people in Wisconsin, of whom more than a few alleged that they were being asked to subsidize a pair of reactors whose primary function was to produce electricity to transmit out of state. Besides, the $1.2 billion for Koshkonong was higher than Wisconsin's total industrial output in 1973, and Koshkonong was nuclear. The public was discovering that nuclear power in Wisconsin, as elsewhere, was more expensive than their proponents had claimed it would be—that the reliability and output of nuclear plants had fallen far short of their promises. And the news media seemed to be printing and airing more and more stories about the dangers and risks associated with nuclear power.

The public had anxious questions to ask; but the utilities in Wisconsin seemed aloof, secretive, and unresponsive. This was a perfect climate for Mike Cherry to exploit when he came into Wisconsin during the summer of 1974 to represent Concerned Citizens of Wisconsin. One of the founders of the group was Dr. Walter D. Moritz, who, like his counterpart George Boomsma in Pennsylvania, had moved away from a city (Chicago) to a rural setting (Fort Atkinson) seeking the good life. Dr. Moritz built a model home for his young family within sight of Lake Koshkonong, established a prosperous local practice as an orthopedic surgeon, and woke up one day to find all of his well-planned dreams shadowed by the utilities' plan to build nuclear-power reactors next to his property. Like many others in a similar position, Dr. Moritz developed an instant interest in the pros and cons of nuclear power. What he learned made him increasingly unhappy; he was determined to oppose Koshkonong. "This plant would be located across the road from me," he said. "That's instant religion."

Dr. Moritz and other local physicians and farmers in the area—unhappy because the Koshkonong plant would usurp 1400 acres of choice crop and dairy land, suck up to 50,000 gallons of water a minute from the shallow lake, and plunk huge radioactive fuel cores in their midst, which could ruin their land in the event of a major nuclear-plant accident—were the original members of Concerned Citizens of Wisconsin. One of their first moves, when they still had less than 50 members, was to hire Cherry as their lawyer. Cherry, convinced that the Koshkonong case might turn out to be a landmark in the nuclear-power controversy, induced Friends of the Earth to intervene on the side of Concerned Citizens of Wisconsin.

Cherry, by this time, was absolutely adamant that nuclear fission power was too dangerous to allow, "particularly," as he alleged, "in the hands of a self-righteous government-nuclear industry, which believes it has the right to lie in the public interest." And he was becoming increasingly frustrated by the concessions that he and other critics had been able to wrest from the AEC and the power companies—which added millions of dollars to the cost of nuclear power, improved its safety, mitigated its impact on the environment, but failed to stop it. Cherry was hungry for total victories. He wanted to stop the construction of nuclear plants, and to impose a moratorium on those in operation.

Before Cherry arrived in Wisconsin, the last public hearing on a permit for a nuclear-power plant before the PUC had lasted just about an hour; the time before that, a similar hearing had wound up in less than a day. Cherry turned this precedent on its head in Wisconsin. And the hearings before the PSC on the Koshkonong reactors, which began in August 1974, had hardly begun to run their course by the winter of 1975. Cherry had tied the whole thing up.

The PSC hearings on Koshkonong, after starting in Fort Atkinson, really got underway in Janesville, appropriately, in a small yellow all-electric demonstration kitchen in the basement of the Rock County Courthouse. Cherry got Wisconsin Electric Executive Vice-President Sol Burstein on the witness seat, where Cherry began to bombard the white-haired utility executive with hard-to-duck questions. At times it was painful to watch. Cherry went after Burstein with a bombastic mixture of calculated outrage, contempt, humor, disdain, charm, and informed invective.

Gerald Charnoff, representing the utilities, kept objecting to Cherry's manner. Charnoff, one of the top nuclear lawyers in the country, was with the firm of Shaw, Pittman, Potts, and Trowbridge, in Washington, D.C. The tall, lanky Charnoff kept complaining that his pugnacious curly-haired adversary was "hysterical" and "playing to the press."

But the press seemed to love it; the 35-year-old

Cherry was colorful copy; and with Cherry making headlines around Wisconsin, the public began to crowd into the small hearing room. One elderly couple stopped Cherry in the hall to tell him, "Until you came here, we never thought the little fellow would ever have a chance against the utilities in this state."

Concerned Citizens of Wisconsin's membership went up from 50 to more than 3000 by the fall of 1974 and was still growing. A candlelight march against nuclear power was organized to converge on the state capital at Madison, a city of 175,000 which became the first municipality in the United States to officially oppose a nuclear plant when a majority of the city council went on record against Koshkonong—saying they didn't want a reactor within miles of Madison, and were against nuclear power in general on the grounds that it was a dangerous and unreliable way to produce electricity.

Governor Patrick Lucey voiced some doubts about the plant. Candidates for local and state offices who said they were in favor of Koshkonong went down to defeat in the November 1974 elections. County boards around Koshkonong passed resolutions condemning the reactors. The *Wisconsin State Journal,* the largest-circulation newspaper in the state, was initially hostile to Cherry and the critics, suggesting they were uninformed agitators who were trying to block progress; but the paper then made an about-face to join the *Milwaukee Journal* and *Madison Capital-Times,* both of which were editorializing against Koshkonong. And the *Milwaukee Sentinel,* another major paper in the state, also played a leading role in the media blitzkrieg that converged on the Koshkonong hearings.

All this put unrelieved pressure on the Wisconsin PSC, which felt it was being formidably squeezed between a vocal public and the powerful utilities. PSC Chairman Bill Eich, who hadn't smoked for years, was going through a pack and a half of cigarettes a day by the fall of 1974. During breaks for lunch and recesses at the hearings, Cherry—a former yeshiva student who can keep spinning talmudic-sounding legal arguments when he's in the mood—would stop Burstein in the halls of the courthouse to insinuate that there was something immoral about the aging utility man's commitment to nuclear power. And when Cherry found it difficult to make certain points during the hearings, he would put them in the public ear by holding press conferences on the steps of the courthouse for radio, television, and the newspapers.

Sol Burstein seemed very patient, calm, and tolerant as the hearings proceeded, which certainly must have been frustrating for him. Every now and then, when Cherry would catch him with a particularly unexpected, telling query, Burstein would take off his glasses, peer thoughtfully at the attorney, and smile thinly, as if to convey a kind of grudging admiration for Cherry's audacity and damaging cleverness.

Wisconsin Electric and its utility allies "were sandbagged" at the PSC hearings, in the words of one of Cherry's admiring supporters. But Cherry's campaign, and ability to command attention and support, were built on facts rather than sand. Right from the start, he was able to thrust the issue of nuclear safety into the middle of the hearings—and get away with it. This was unprecedented, anywhere, on the state regulatory level. Attorney Gerald Charnoff objected, contending that the courts had upheld the AEC's right to be the sole judge on matters of nuclear safety. But Cherry was able to convince the PSC hearing examiner that the safety issue lay at the heart of the debate on nuclear reactor economics and reliability, which would have to be considered by the PSC because it had been charged with the responsibility of assuring the public reliable electricity at a reasonable cost.

Later, on May 1, 1975, the Wisconsin PSC commissioners, by a 2–1 vote, ruled that the commission had the jurisdiction to review all safety and reliability problems of nuclear-power stations if such problems "substantially impair the efficiency of the service of [a] public utility."

With Charnoff objecting, and Sol Burstein looking wary, Cherry paraded a host of damaging facts about the dangers of nuclear power, contending that the safety issue was responsible for the fact that nuclear-power plants on the average were producing at less than 55 percent of capacity.

Then Cherry questioned Burstein closely on the utility claim that southern Wisconsin would need 59 percent more electricity by 1981 because of the "population and industry" Burstein said was moving into the state. The PSC, which had hired a number of bright, young, well-educated, reform-minded staff members under Chairman Eich, also began to question the growth of electricity predicted by the utilities—with Cherry pointing out that conservation-minded Wisconsin consumers, trying to hold down their electric bills, had made sharp inroads into the growth of electrical consumption.

But the thing that really seemed to turn the public on about Koshkonong, and against it, was when Cherry squeezed the fact out of Burstein that the utility consortium had spent $11 million on the Koshkonong project before the hearings began—before the PSC had given any approval for Koshkonong; that the utilities had signed contracts to spend an additional $23 million on the plant between September 1, 1974, and September 1975, and had advanced $2 million to the AEC on a contract for enriched uranium fuel, which meant that the utilities were prepared to spend millions of dollars without actual assurance that the PSC would give them a construction permit. Many people began to cry that this would put tremendous financial pressure on the PSC to approve the project, even if it had misgivings, just because there was already so much money invested in the Koshkonong nuclear plant.

The PSC, on its own behalf, said it resented this pressure, than handed down an unprecedented temporary injunction, perhaps the first of its kind anywhere, prohibiting any more prior spending on Koshkonong. This was later rescinded, but the utilities were ordered to file monthly public reports with the PSC showing how much was being spent on Koshkonong so that the utility shareholders would know they were proceeding at their own risk. And the critics began to say that the shareholders, rather than ratepayers, should be compelled to pick up the multimillion-dollar tab in the event that the Koshkonong nuclear plant wasn't approved.

After this, Cherry was able to grab more headlines when he disclosed that the Wisconsin utilities had signed a letter of intent to buy six reactors from Westinghouse. Cherry presented further evidence to indicate that the utilities were ultimately planning to build six reactor units rather than two at Lake Koshkonong or elsewhere in the state. Prior to these disclosures, the PSC was essentially ignorant of the utilities' expansion plans. This was a startling revelation which enhanced a feeling on the part of the public that the "utilities think they can do anything they damn please without telling anyone what they're up to," in the words of one disgruntled farmer.

By the fall of 1975 Wisconsin Electric cited a slowdown in electric growth and uncertainties about government approval in its decision to postpone construction of the Koshkonong reactors by at least two years, adding that the estimated $1.2 billion cost of the project had gone up by 24 percent since the summer of 1974. In the meantime William Eich stepped down from the PSC to accept a state judgeship. He was replaced by Matthew Holden, Jr., a black professor of political history at the University of Wisconsin, who said he was a "nuclear skeptic," sympathetic, perhaps, to the idea of a moratorium on the construction of nuclear-power plants.

Friends of the Earth, which supported the intervenors in the Koshkonong hearings, was immensely pleased by developments there. "We need five or six more Mike Cherrys around the country" is the way the organization's energy projects director (and nuclear specialist) Jim Harding put it in 1975, "to make the regulatory agencies work a lot harder. Because nuclear power will probably be stopped locally before it's stopped nationally."

About the same time Mike Cherry was insisting that

state regulatory agencies have the right to review nuclear-plant safety factors because of their potential effect on electric rates, the folks who work for Ralph Nader's Citizen Action Group on Capitol Hill were organizing the first national conference against nuclear power, which was held the weekend of November 19–20, 1974, in Washington, D.C. Almost 900 delegates from 39 states arrived to mill around the hallways of the Statler Hilton Hotel, exchanging information and tactics, crowding into busy workshop sessions, and moving into the presidential ballroom to hear a host of prominent speakers denounce nuclear power.

Chauncey Kepford from York and George Boomsma from Lancaster were there, mingling with representatives from Concerned Citizens of Wisconsin and hundreds of other antinuclear grass-roots groups. Many prominent names were in attendance from the national organizations, such as the Union of Concerned Scientists, the Committee for Nuclear Responsibility, Friends of the Earth, Task Force Against Nuclear Pollution, National Intervenors, and the Natural Resources Defense Council.

Ralph Nader and his team called the gathering Critical Mass '74—a takeoff on the amount of nuclear material necessary to trigger a spontaneous nuclear reaction—and said its purpose was to "provide a national focus on the risks and consequences of nuclear fission and expand the citizen base . . . to move toward a nuclear-power moratorium." Nader also coined the conference theme: "A nuclear catastrophe is too big a price to pay for our electric bill." Nader workers Joan Claybrook and Faith Keating, among others, put together a heavy, impressive 161-page booklet, *Citizen Manual on Nuclear Power '74,* packed with facts about the dangers and shortcomings of nuclear power and tips on how to oppose it, which was handed out to the delegates.

Anthropologist Margaret Mead and antiestablishment journalist I. F. Stone were among the celebrities who came to urge the antinuclear troops to step up their campaign against reactor power plants. Actor Robert

Redford was there too, saying he wanted to learn more about nuclear power; he later indicated that he had grave misgivings about it.

Nader said, "It is clear that there are a number of independently sufficient grounds to call for a moratorium, and the cumulative impact is overwhelming." He warned against the dangers of nuclear-plant accidents or sabotage; noted that a "garrison state" might be erected in an attempt to guard plutonium from would-be bomb makers; said the economics of nuclear fission power depended upon government subsidies; and urged the practice of conservation and if necessary the use of pollution-controlled coal until crash programs could be developed to produce new sources of energy —among them solar and geothermal power—before the turn of the century.

Seven prominent scientists wound up the Nader-called conference with the signing of a petition urging Congress to appoint a special committee to investigate the impact of nuclear fission power on the health, safety, and security of the nation. They were Nobel laureates Hannes Alfvén and George Wald; former AEC researchers Dr. John Gofman and Donald Geesaman; physicist Henry Kendall; Barry Commoner, a biologist-writer from St. Louis, one of the pioneer critics of nuclear power, and chairman of the Scientific Institute for Public Information, which alleged that the AEC had tried to minimize the dangers of nuclear power while downplaying the potential of solar energy; Dr. Dean E. Abrahamson, physicist and physician, professor of public affairs at the University of Minnesota, and one of the important early critics of nuclear power; and last, nonscientist S. David Freeman, engineer and lawyer, and former director of the White House Office of Science and Technology during the Kennedy, Johnson, and Nixon administrations.

Senator Mike Gravel said, "One important consequence of Ralph Nader's antinuclear conference will be recognition in Congress that grass-roots opposition to nuclear power has become a political force to be reckoned with." The Atomic Industrial Forum, reporting on the conference, came to the same conclusion, saying, "The antinuclear movement is now a full-fledged political movement. . . . More grass-roots work has clearly been completed than most industry observers had realized."

By the summer of 1975 Ralph Nader had institutionalized the antinuclear movement with the formation of Critical Mass, an organization named after the successful protest conference in November—with hundreds of affiliates and some 200,000 supporters; hundreds of thousands of people around the nation were signing a national Clean Energy Petition sponsored by the Task Force Against Nuclear Pollution; and the Sierra Club, Common Cause—the political watchdog group—and the 100,000-member Wilderness Society were calling for a moratorium on the construction of nuclear plants.

All this citizen pressure was converging upon Congress. But when members of Congress looked to the scientific community for guidance, they found it was becoming increasingly polarized over the issue of nuclear power. In January 1975, for instance, a group of 32 prominent scientists, including 10 Nobel prize-winners, issued a manifesto endorsing nuclear fission power. Next, on August 6, to commemorate the 30th anniversary of the bombing of Hiroshima, a statement signed by 2300 scientists, engineers, and physicians was delivered to the White House, warning of the dangers connected with nuclear power, urging a slowdown on the construction of reactor plants, and calling for urgent programs to develop alternate sources of energy.

The pronuclear proclamation was drafted by Dr. Hans A. Bethe, a world-famous Nobel laureate in physics, and Dr. Ralph Lapp, a physicist, consultant, and writer. It warned that the "republic is in the most serious situation since World War II. Today's energy crisis is not a matter of just a few years, but of decades . . . the high price of oil which we must now import in order to keep Americans at their jobs threatens our economic structure." These scientists said "conservation is essentially the only energy option" for the immediate future and that "we shall have to make much greater use of solid fuels. Here uranium and coal are the most important," adding that "we can see no reasonable alternative to an increased use of nuclear power to satisfy our energy needs" because "the benefits far outweigh the possible risks." Most of these thirty-two scientists, perhaps not surprisingly, who signed this statement had close ties with the nuclear establishment.

The other declaration was initiated and circulated by Dan Ford and Henry Kendall of the Union of Concerned Scientists. Among the 2300 signers were a number of names not normally among those of the antinuclear critics, such as James B. Conant, president emeritus of Harvard; George B. Kistiakowsky, science adviser to President Eisenhower; former MIT physics chairman Victor Weisskopf; Nobel prizewinning physicist Julian Schwinger; and Richard F. Post, deputy associate director of fusion research at the Lawrence Livermore Laboratory in California. This statement noted that "nuclear fission creates massive amounts of radioactive by-products, posing grave potential hazards," with the result that an "early optimism" for nuclear fission power "has been steadily eroded as the problems of major accidents, long-term radioactive waste disposal, and the special health and national security hazards of plutonium become more fully recognized."

The declaration also said "the problems now besetting nuclear power are grave, but not necessarily irremediable, [but] the country must recognize that it now appears imprudent to move forward with a rapidly expanding nuclear-power-plant construction program. The risks of doing so are altogether too great. We therefore, urge a drastic reduction in new nuclear-power-plant construction starts before major progress is achieved in the required research and in resolving present controversies about safety, waste disposal, and plutonium safeguards."

The signers went on to urge the stepped-up use of coal, conservation of energy, and a "full-scale research and development effort to create more benign energy producing technologies that can make use of the energy of the sun, the winds, the tides, and the heat in the earth's crust. Fusion energy research should also be given an enhanced priority." The declaration ended on the note that "it was a serious error in judgment . . . to devote resources to nuclear development to the virtual exclusion of other alternatives."

Some critics began to hope for changes in the Joint Committee on Atomic Energy. Congressmen Chet Holifield and Craig Hosmer from California, both influential committee boosters of atomic energy, retired after 1974, leaving a power vacuum. Reports in the press indicated that Senator John O. Pastore, a Democrat from Rhode Island and the new chairman of the joint committee, might be inclined to give careful consideration to both pros and cons of nuclear energy.

In a brief meeting with Senator Pastore in May 1975, he gave me some of his views on the subject. "I'm a staunch advocate of nuclear power," he said, "because it's the only practical, competitive source of energy in the near future. But I've also insisted that it has to be made safe." Pastore added that the antinuclear movement had a "salutary effect. This opposition has been building for a long, long time. There's a tremendous amount of emotion in back of it. And this aroused public has helped to make the manufacturers more conscious of the safety issue, and of their responsibility to it—and strict safety measures are an absolute necessity. But once we come to the end of the road, I think that most members of the Congress will conclude that nuclear power is essentially competitive and safe."

The nuclear establishment, reacting to the increasing threat of the antinuclear movement, said it was going to step up its efforts to woo public opinion and members of Congress over to the pronuclear side. The Atomic Industrial Forum, the industry's "trade association," eager now to actively lobby, set up headquarters in Bethesda, Maryland, where the NRC's main office was located. A new group called the American Nuclear Energy Council, headed by former Congressman Craig Hosmer of the Joint Committee, established its office next door to the NRC office in downtown Washington, D.C. Hosmer said, "People have been listening to the more vocal opponents for too long. Now it is time to get out and tell people that nuclear power is OK; that the risks of nuclear power are minimal in relation to its benefits."

Physicist Hans Bethe and retired Navy Admiral E. R. Zumwalt, Jr., came forward to head another new pronuclear group, called Americans for Energy Independence, headquartered in Arlington, Virginia, but Zumwalt later dropped out.

In August 1975, a group of engineers and lawyers who worked for the utility companies and belonged to an organization called the Council on Energy Independence, released the findings of a Harris poll reporting that 63 percent of those surveyed were in favor of building more nuclear plants. But this sentiment, real or illusory, could be temporary, because the citizens' movement against nuclear power was getting stronger, and spreading. This was apparent when more than a thousand critics from around the country came to Washington, D.C., on November 16–17, 1975, for Critical Mass '75. *Weekly Energy Report,* which devoted a special issue to the second annual national conference called by Ralph Nader, said, "More and more the antinuclear movement is behaving like the lions." The publication described critics as "tighter, better organized and better focused . . . and, perhaps most important, they believe they are winning."

Ed Koupal from California, the home of the nation's first nuclear initiative, a strategy which had attracted supporters in 20 states for a nuclear-power moratorium, called the gathering a nonpartisan "political convention" and said, "We're going to beat this [nuclear] industry, and we're going to beat it to death."

Nader pointed out that "the electric utilities have canceled or postponed more than two-thirds of the previously announced nuclear plant construction projects in the U.S. . . . in the last two years alone. . . .

New orders for nuclear plants have evaporated as well. The statistics make it plain that nuclear power is the Edsel of energy technology: it just won't sell anymore." He warned attempts would be made "that we socialize the losses and risks of a failed technology." Nader then struck Critical Mass '75's major theme: "The highest priority of the citizen movement must now be devoted to stopping the Ford administration's nuclear-power bailout program." This would include the $100 billion Energy Independence Authority proposed by the Ford administration and several other billion-dollar subsidy programs.

David Comey underscored Nader's concern at the conference by calling the taxpayer "the ultimate bailout mechanism in our system of 'lemon socialism' in which free enterprise gets to keep the plums and the government takes over the lemons."

Llewellyn King, the editor of *Weekly Energy Report,* summed up the new direction of the antinuclear citizens' movement in this way: "It is now fighting the nuclear power in the political arena. The courts and the licensing are only a small part of the battle today. The plans that were being drafted in Washington this week [of the conference] are plans for a political victory."

But the nuclear establishment, with its $100 billion investment and its future to protect, was still full of fight, as it demonstrated in December 1975 when both the House and the Senate—though there was considerable debate, and an increasing show of antinuclear strength—voted to extend the controversial Price-Anderson Act regulating nuclear-plant insurance for another ten years. This pronuclear victory was destined to give brief comfort to the nuclear advocates, because 1976 looked like it was going to be a very bad year for them.

The nuclear establishment's reputation was severely tarnished, to say the least, by the February 2 resignations of three management-level engineers from General Electric's nuclear reactor division, who said their long experience in the field had convinced them that nuclear power was a profound threat to the future of man.

Next, on February 9, project manager Robert D. Pollard of the Nuclear Regulatory Commission said he was resigning to protest the dangers of nuclear power. Pollard singled out Consolidated Edison's nuclear power complex on the Hudson River, 26 miles north of Manhattan, saying, "The Indian Point plants have been badly designed and constructed and are susceptible to accidents that could cause large-scale loss of life."

One critic told me these defections from the middle-management ranks of the nuclear establishment were a "sure sign that the ship is beginning to sink. More and more people are going to get off. They're just trying to figure out how."

I'm sure a number of observers had the same thought. One thing was sure. In the space of four short years, from 1972 to 1976, the tables had turned. And the powerful nuclear establishment, which had once seemed omnipotent, was beginning to run scared. Suddenly, in some circles, as if overnight, the word "moratorium," which only yesterday had an avant-garde sound to it in regard to nuclear-power plants and their related facilities, now sounded almost passé. And one could hear the word "phaseout" instead.

IX
The Future

The antinuclear critics have always insisted there are
a host of safe, productive alternatives to nuclear power
which we could develop for widespread use within a
relatively short period of time—if we set our minds to
it. These include conservation, in terms of more efficient use of energy, which could bring immediate divi-
dends; solar power; wind power; converting garbage
and waste into several energy sources; oceanic thermal
power; and geothermal energy.

While I was in California I decided to take a look at
geothermal power, and found it 90 miles north of
San Francisco, at the remote end of a winding, twisting,
scenic road that climbs up through the rugged hills of
Sonoma County where, in a vast valley, fields of clouds
of billowing white steam come streaming eerily up
through cracks in the gray earth. The Indians thought
it was a place where the spirits dwelled. William B. Elliot, an explorer-surveyor, was out hunting grizzly bears
in 1847 when he smelled the sulfur and saw the hot
springs and the wafts of steam; he was afraid he had
discovered the "Gates of Hell." What Elliot had actually
stumbled upon was one of the few places in the world
where molten rock, called magma, still cooling from the
creation of the earth, is fairly close to the crust, where it
gives off gas and water vapor that mix with surface
water to create hot springs and steam.

By 1975 a three-company group, led by Union Oil,
was turning the Big Geysers area into the world's largest
geothermal energy project, where giant bits go digging

down to depths of 2 miles and more to drill wells that approach this mass of molten rock to tap steam and send it forth to nearby turbine power plants that generate 500 megawatts of electricity—enough to supply a city of 500,000 with all the power it needs, at less than half the cost of using fossil or nuclear fuels.

A number of scientists claim that geothermal power could supply us with 30,000 megawatts of electricity by 1985. That's the same amount nuclear fission reactor plants in America were supposedly supplying in 1975—after 30 years of development and an investment of over $100 billion. They say geothermal power could deliver 200,000 megawatts by the year 2000— enough to supply the United States with 20 percent or more of the electricity we might need by then. Advocates of geothermal power added that its environmental drawbacks, which are relatively slight, are virtually nonexistent when weighed against the potential dangers inherent in nuclear power.

Most of the molten magma to supply geothermal energy is 20 or more miles below the crust of the earth; in some places it is 5–10 miles down. Wells would have to be drilled, into which water can be pumped to create hot water and steam to produce electricity. Some scientists believe there is enough geothermal energy under California's Imperial Valley alone to supply us with as much electricity as the whole nation consumed in 1970.

Developing new sources of power, such as geothermal energy, has been tied to a fast-rising national awareness that we have been squandering our irreplaceable supplies of oil and natural gas as if tomorrow would never come. But it did—and abruptly, with the Arab oil embargo of October 1973, when the alleged machinations of the major oil companies had motorists lined up at service stations from New York to California, unhappily watching the price of gas go up.

With the future of nuclear power somewhat in doubt, interest in other energy options was increasing, although slowly, it seemed, in the ranks of the federal government, which until 1974 acted as if the development of nuclear fission power was the only energy goal worth pursuing. By 1975 this one-sided view was changing—but not by so very much, as this budget proposed by ERDA for fiscal 1976 indicates: fission, $763 million, or 45 percent of the budget; fossil fuels, $417 million, 25 percent; fusion, $264 million, 16 percent; solar, $89 million, 5 percent; geothermal, $32 million, 2 percent; other energy research, $44 million, 2 percent; and conservation, $73 million, 4 percent. Some critics who looked at this budget complained that nuclear fission power was not only unsafe, but also incapable of meeting our energy needs on any scale that could be termed reasonable.

In 1974, for example, the prestigious Club of Rome, weighing our dwindling global resources against a leaping growth in population, published its second commanding report, *Mankind at the Turning Point*. The report attacked the concept that nuclear reactors—in particular the liquid-metal fast-breeder model—could provide a panacea for our future energy needs, as its advocates, including those in AEC, were claiming it could. After weighing the potential electrical producing capacity of the breeder, the Club of Rome authors concluded that the world would need 3000 nuclear parks, each of them with at least eight breeder reactors, to produce enough electricity to meet global demands 100 years from now. This would mean building four reactors every week between now and then, the report said, adding that this figure did not include the prospect of building two reactors per day to replace those which wear out after an average of 30–40 operating years.

In the United States alone, according to the Club of Rome—which based its analysis on AEC predictions of nuclear growth—relying solely on atomic energy would "require more than 50 major nuclear installations on the average in every state in the union." This would mean producing and handling millions and millions of pounds of plutonium. There would also be the awesome burden of trying to safeguard these 3000 nuclear parks from the dangers of accidents, sabotage, theft, and war, which could release enough plutonium into the environment theoretically to poison the earth for hundreds of thousands of years.

These were some of the reasons why the Club of Rome report said nuclear power was unacceptable and urged us to develop solar power. In reality, however, we have always been using it; scientists tell us that every bit of life and energy on earth is running on sunlight: it is stored in coal, natural gas, and oil, in wood and plants and garbage; it causes the atmosphere to circulate, creating the wind, the waves, and the tides and currents of the sea.

We put up windmills to capture it; dam the rivers to spin our turbine-generators; burn the hydrocarbons —in wood, coal, natural gas, and oil—to release stored solar energy to heat our homes and produce electricity.

Wood used to be our primary source of energy; 80 percent of energy came from wood around the time of the U.S. Civil War. By 1910, 79 percent of our energy was coming from coal; this in turn gave way to natural gas and oil, which were supplying us with 75 percent of our energy by the 1970s.

Natural gas and oil seemed cheap, relatively easy to find, process, and exploit; for a long time many people wanted to believe the fossil fuels were virtually inexhaustible. There were calls for conservation—for more efficient use of these fuels; but they were largely ignored. Instead, the oil and electric companies kept urging us to consume more and waste more fuel and electricity. From 1946 to 1968, while the national population went up by 43 percent, the consumption of electricity rose 436 percent.

Oil and natural gas comprise only about 11 percent of the world's fossil-fuel base—formed over aeons by the decay of plants and animal life trapped deep beneath the earth and subjected to tremendous pressures and heat. Now the experts are telling us that these short-term fuels, at the present rates of consumption, will be exhausted by the year 2050—give or take a few decades. In any case they are going fast; as they do, their costs will continue to soar.

We are now using over 6 billion barrels of oil a year,

which is why people such as Friends of the Earth President David Brower keep urging us to conserve energy. Certainly there is plenty of fat that could be trimmed from the energy bone. Many cars are only getting 7 miles to the gallon; the tiny pilot lights on our kitchen ranges are consuming up to a third of the gas our stoves use.

The Ford Foundation-funded Energy Policy Project, in a major study, recommended in 1974 that we could cut the growth of electricity in half without hurting the national economy; it urged the federal government to take the lead in a national drive to conserve energy by making more efficient use of it.

Energy efficiency certainly seems to be one of the waves of the future. In fact, by the time 1975 was rolling into 1976 there were so many studies, suggestions, and logical-sounding programs dealing with this topic that it was difficult to keep track of them. The American Institute of Architects, for instance, brought out two reports in 1975 describing a program for energy-efficient buildings which, if acted upon today, would save us around 12 million barrels of oil per day in 1990; the benefits, according to the AIA, would be almost immediate—a saving of 750,000 barrels of oil per day after the first year. Installing proper insulation and other energy conservation measures in existing buildings could result in energy savings of 30 percent, according to the AIA, while new construction that incorporated energy efficiency concepts and materials could bring in energy savings of up to 60 percent.

The technology to implement energy efficiency buildings is available today, the AIA said, stressing that a program of this sort would create jobs to improve the economy and our health and comfort at the same time. Another important point, according to the AIA, is that capital used to conserve this energy would equal "as much energy as the projected 1990 production capacity of any one of the prime energy systems: domestic oil, nuclear energy, domestic and imported natural gas, or coal."

Another report, called "Assessing the Potential for

Fuel Conservation," prepared for the Institute for Public Policy Alternatives by Drs. Marc Ross and Robert Williams, said that "over 60 percent of the potential savings lie in four areas: space conditioning, the automobile, industrial congeneration of steam and electricity, and commercial lighting." Ross and Williams said the consumption of auto energy, for example, could be cut by one-third with "optimization of engine size, a 20 percent total weight reduction, radial tires, and modest streamlining."

Dow Chemical coordinated a study on the potential of industrial steam for the National Science Foundation. It reported that large industries that generate their own steam could convert about 45 percent of this heat that's now wasted into electricity for sale to utilities. This could be accomplished, the study said, by adding turbines and related facilities to large industrial plants at relatively low cost; and the report claimed that this utilization of steam heat could produce an amazing 71,105 megawatts by 1985, which would represent about half the power American industry will reportedly be demanding that year.

Lee Schipper, a physicist with the Energy and Resources Group of the University of California at Berkeley, who addressed the Critical Mass '75 conference, summed up the energy efficiency ethic when he said, "Conservation does not mean doing without, but doing better. . . . Aside from the changes in cars and in added recycling, we'll hardly notice the effects of conservation—except at the bank, for in addition to saving millions of barrels of oil each day we'll save billions of dollars a year, billions that can provide a higher standard of living. Equally important are the indirect benefits of using energy more efficiently. Greater employment results from increased energy productivity, with more workers needed to build and run a more sophisticated, more efficient industry, and more labor needed to make well-designed, insulated buildings and homes, efficient appliances and autos, and mass transit."

Many people were also beginning to find coal—of

which we have at least a 400-year supply in this country alone—more attractive than it seemed a few years ago because of the environmental hazards associated with mining and burning it. In 1973 Secretary of the Interior Rogers C. B. Morton was saying, "Coal could supply our energy needs for hundreds of years. . . . Clean liquid and gaseous fuels can be made from coal; electricity can be generated from coal in an environmentally accepted manner, provided the present development program is carried through to a successful conclusion."

By September 1974, after coal began to receive more research support, the Environmental Protection Agency was announcing that there was enough evidence and experience to indicate that filter devices called scrubbers could be installed in boilers and smokestacks to remove dangerous pollutants from unmined high-sulfur coal in the East, which in turn should rule out proposals to strip-mine the West for low-sulfur coal.

Jim Harding, of Friends of the Earth, told me that technologies have been pioneered in Great Britain to generate the widespread use of heat instead of electric power from coal. These look promising, Harding said, because "fluidized bed boilers can be built and operrated in urban areas—piping heat directly to commercial, industrial, and residential users. This would result in great cost savings over electric conversion, and with very, very few pollutants because of the extremely sophisticated air-pollution-control abilities of the fluidized bed boilers," Harding said.

We know that much of the energy involved in electricity is lost. S. David Freeman, who headed up the 1974 Ford Foundation energy study, says, "The conversion of primary fuels to electricity always involves a substantial energy loss. In the process of electrical generation and transmission, about 65 percent of the energy content of the fuel is lost."

Harnessing the direct rays of the sun could save us at least a million gallons of oil a day by 1985, according to President Ford and Federal Energy Administration head Frank Zarb; both of them declared in May 1975 that solar energy was going to play an important role in their goal of trying to reduce petroleum imports by 3–5 million barrels a day. ERDA also announced that it had a program which would enable solar power to supply the United States with a quarter of its energy by 2020.

These goals seemed extremely modest to scientists who have been endorsing solar power for years, and who have repeatedly pointed out that sunshine is safe, clean, and inexhaustible. In 1970, for example, the late Dr. Farrington Daniels, one of the world's leading chemists, was saying that "solar energy is amply adequate for all the conceivable energy needs of the world." Other scientists said that only 1 percent of the sunshine that fell on the United States was enough to supply us with all the energy we would ever need. Solar power is available right now. Although it is a poor source of energy for electricity at present, this is not considered a weakness because most of our real energy needs do not require electricity. More than 300 homes, offices, and schools around the country in 1975 were using the direct rays of the sun for their heating and cooling. The most immediate practical application of solar power will probably continue to be for individual buildings, which can be of tremendous importance in a country like the United States, where residential use accounts for about 34 percent of all the electricity we consume. The technology to use solar energy for heating—and sometimes cooling—is ready now and being improved all the time.

One method is to use panels of glass, usually slanted, sometimes orbiting to track the sun, on the roofs or sides of buildings to catch the rays of the sun, which pass through the solar panels into a box with a black bottom to retain this energy, which, in turn, heats water or air that is piped through the box for immediate use or storage. Another approach is to heat tanks of water on the top of buildings by the sun to furnish hot water and heating. Usually these tanks are covered by panels that slide open in the day to capture the heat from the sun and close at night to retain the solar energy. In winter the hot water radiates heat through the ceiling;

in summer the process is reversed for cooling—the panels are covered during the day to turn away the sun, and open in the evening to let the night air cool the water. This method works especially well in the Southwest, where the days are hot and dry and the nights are cool.

There is also considerable interest in the use of photovoltaic cells, which charged our spaceships with energy to use during their trips to the moon. These cells are usually made of silicon or cadmium, which convert sunshine directly into electricity. Photovoltaic cells could be installed in modular units—consisting of a network of tiny cells—in power plants, batteries, and on rooftops or other parts of a building. Photovoltaic cells have two great advantages: they can function on cloudy days as well as in direct sunlight; and they can be replaced without disrupting the whole modular system. But they are terribly expensive, and they will not be available for widespread use until mass production techniques are developed to greatly reduce their cost.

But scientists claim that enough interest and sufficient funding could begin to make this and other forms of solar power commercially attractive and available on a widespread basis between 1985 and 2000.

The pursuit of direct solar power could also take the form of acres of mirrors in the deserts of the Southwest to capture the sun's rays, convert them into thermal energy, and send the heat to a central station for storage in molten salts to produce steam to drive a turbine generator. This also would be prohibitively expensive today. Many square miles of land would probably be required for the solar collectors; and the technology to transport the heat over great distances to areas where it is needed is still in its infancy.

Other solar farms have been contemplated that would involve the cultivation of large acreage of crops, trees, algae, cane, or seaweed that could be harvested and burned to produce heat energy for direct use or conversion to electricity. This, in essence, would short-cut the process which gives us heat energy from fossil fuels. In back of all our energy is the process of photosynthesis, wherein green plants absorb sunlight to manufacture carbohydrates—sugars, starches, and cellulose —which, as their name suggests, contain atoms of carbon and hydrogen. These are transformed over millions of years into hydrocarbons in coal, natural gas, and oil. But we know from burning wood that we don't have to wait aeons to use carbon and hydrogen for energy; we can have it in a matter of years, months, or even days in the case of algae.

This brings us to methane gas, another enticing new energy source. County farm agents in Pennsylvania, for example, working with Pennsylvania State University extension campuses, have been demonstrating the workability of homemade manure gas-generating kits. Any farmer with 15 head of cattle would have enough animal waste on hand to produce clean methane gas for all his farm equipment, his trucks and car, and to heat and cool his home. The remaining sludge retains its high nitrogen content that can be returned to the soil as fertilizer, so important to a world stalked by famine. Some urban homeowners, instead of throwing their garbage away, have been converting it into methane to heat their city homes. Manure and garbage can also be turned into low-sulfur oil—enough perhaps for more than 2½ billion gallons a year in this country.

The carbon and hydrogen in waste and garbage can also be converted into methanol, or wood alcohol, the clear fluid that we find in Sterno and use to heat food or keep it warm. All sorts of vehicles were operated on wood alcohol in France and Germany during both world wars when gasoline was scarce. Many racing car drivers have preferred alcohol to gasoline for some time; there is every reason to believe that ordinary cars could be modified to make direct use of methanol as a fuel or as a blend or additive with gasoline. Right now methanol is expensive, but if demand increases, widespread manufacture could drive the current price way down. In fact, methanol seems to have so much fuel potential for vehicles that some nuclear critics, who have been endorsing it as one of the many alternatives to nuclear power, are afraid it will encourage a continued reliance upon the automobile, which they deplore for a host of social, energy, economic, and ecological reasons.

Converting garbage, refuse, and waste into an asset, in the form of methanol fuel, has tremendous implications for our urban centers. The National League of Cities and the U.S. Conference of Mayors, in a joint report released in 1975, warned that nearly half of our cities would have no place to put their garbage in a few short years, which is one of the reasons why many cities are beginning to build plants to either reclaim fuel from waste or burn it directly for power. Methanol, which burns with a clean, pure flame, can also be used in power plants without polluting the atmosphere. Today most of the methanol in this country is manufactured from natural gas. It can also be produced from coal, and piped directly out of coal mines in large quantities. Its real potential lies in the fact that it can be manufactured from our vast quantities of waste and garbage.

Some people would like to harness nuclear fusion, if they could. This process would fuse instead of split atoms, namely light atoms of deuterium—found in ordinary seawater—into helium at fantastically high temperatures to release a tremendous flow of energy. Essentially this is the reaction that is responsible for sunlight; it is the source of energy in the stars, and the thermonuclear hydrogen bomb. Fusion is controversial, and becoming more so all the time. Those who believe in it, or profess that it is worth pursuing, are trying to obtain a worthwhile net energy gain from a reaction that takes place at temperatures ranging from 100 to 300 million degrees C. At this extremity, everything turns to positively charged gases called plasma. It has been impossible to sustain and confine plasma at these fantastic temperatures. Recent studies indicate that the rain of radioactive neutrons released by the process might bombard the walls of a reactor to such an extent that they would have to be replaced every five years or less, which would be prohibitively expensive and impractical.

Detractors also warn that fusion would produce other

serious radioactive contamination; that a fusion reactor might explode; and that it would require enormous investments of capital and government subsidies in the pursuit of an absolutistic technology that would be inherently dangerous, undemocratic, impractical, or all three.

William Heronemus, at the University of Massachusetts, is one of the leading boosters of wind power. He says the "entire electricity demand of the six-state New England region could be satisfied by wind power alone by the year 2000 at a cost which even now would be competitive." Heronemus also says that "tidal power possibilities, discarded by the United States in the past but carefully set aside by the more canny Canadians, should be reexamined at once."

The National Aeromautics and Space Administration believes that stringing giant windmills across the country, where they could do the most good, would produce the amount of electricity we consumed in this country in 1970. Heronemus advocates stringing a grid of some 300,000 giant windmills, many stories tall, across the Great Plains to meet our energy needs. Heronemus would also like to put windmills off the coast of New England to convert seawater by electrolysis into hydrogen, which in turn would be piped ashore and stored. Hydrogen, as a fuel, has many of the characteristics of natural gas.

Then there is direct heat from the oceans, covering around 70 percent of the earth's surface, which absorb tremendous amounts of sunlight each day. In the tropic seas, water on the surface may reach a temperature of 80°–90° F; underneath, cool water moves along the ocean floor from the arctic poles to the equator. A floating power plant in the ocean could use this difference in temperature between the warm surface water and the cool water underneath to produce electricity.

Robert H. Douglas, manager of Ocean and Energy Systems Projects, a subsidiary of TRW Systems and Energy, Redondo Beach, California, told a House subcommittee on energy in June 1975 that ocean thermal energy conversion system plants might be highly competitive with oil-fired power plants. He described the system as follows: A large, floating plant that looks something like an oil rig takes warm water from the surface of the ocean and draws it into evaporators, where it converts liquid ammonia into a vapor to spin a turbine. After it passes through the turbine, it passes into a condenser, where the colder water from below cools the ammonia vapor and returns it to its liquid state, where it is pumped back up into the boiling chamber, ready to be converted into steam again to spin the blades of the turbine generator to produce electricity.

Douglas says it would be possible to build and demonstrate an ocean thermal conversion plant for about $500 million.

One of the big issues in the ongoing controversy over nuclear power and alternative sources of energy is sure to be jobs and the economy. Congressman Mike McCormack calls it the "big issue," and predicts that "labor and industry are going to line up behind nuclear energy to make sure we maintain jobs and create more of them. If we did not have nuclear energy available to us for the coming decade," he said in 1975, "the fate of this country would indeed be black. . . . Our standard of living, our economic stability, our political institutions and even the safety of our streets may well hang on whether or not we are able to produce enough energy to maintain the job market. . . ."

Dr. John Gofman has this retort: "In our modern society the last refuge of the promoter is to threaten people with loss of jobs, loss of livelihood, lack of food, and a return to life in the cave." He called this tactic a "brazen promotion of nuclear power." Gofman is among the many, who in his words feel that "energy efficiency, as a concept in conservation, is likely to become the greatest stimulus to the job problem as well as the truly significant way to approach the energy supply problem in the near and midterm. . . . Investment in energy efficiency is far more attractive than investment in wasting energy, which is what choosing the light-water [nuclear] reactor program represents.

"In fact, one of the greatest threats to the economy," Gofman said in 1975, "is nuclear power. One major nuclear accident would throw thousands of people out of work, and throw thousands more into panic. It could disrupt the entire economy. The whole rationale behind nuclear power is tremendously illogical. The thing to do is create more jobs. We need labor-intensive jobs. Moving into the production of wind power alone could provide more than half the jobs that Detroit now offers. And anyone who knows, knows we can begin to build windmills now; and that solar power is viable economically and jobwise too. Solar power is guaranteed to give us net power, and soon."

A number of months after I talked to Gofman in San Francisco, I spent an afternoon with Wisconsin Electric Power Company Vice-President Sol Burstein in his Milwaukee office. I found this executive engineer to be charming, bright, and thoughtful, a man who clearly felt that nuclear power was both safe and desirable; who brooded deeply over the thought that without nuclear power to supply us with electricity we might be courting widespread unemployment and hardship, which Burstein felt could lead to an era of totalitarianism in this country.

The critics, however, and they seem legion, are well prepared to argue an opposite point. They believe that nuclear power, which represents a centralized, energy-intensive economy requiring an enormous capital investment and the monopolistic partnership of big business and big government, is actually taking away jobs and paving the way toward a Big Brother society.

Dr. Henry Kendall, for one, calls nuclear power "the least job-intensive, the most capital-intensive" industry in the country. Addressing the Critical Mass '75 conference, he cited figures for the electric industry as a whole to show that electricity requires about 15 percent of the nation's capital, but provides only about 2 percent of the jobs. Kendall said that 30 percent of the nation's expendable capital would go for huge central power stations by the year 2000 if we kept up the electrical growth rate of the 1970s. This would divert tremendous financial and natural resources from programs vital to society and the health of the economy, according to Kendall, who insisted that "electricity replaces people" from jobs.

Pehraps the one point to remember, in this thicket of many pros and cons, was best expressed by David Brower of Friends of the Earth. I was in the room when he was being interviewed for San Francisco television in 1975. When he was asked about the nuclear-power controversy, Brower hesitated for a moment, to collect his thoughts, ran his fingers through his white hair, and turned his blue eyes toward the lens of the camera. "I think we're going to stop nuclear power because it has to be stopped," he said. "Let's put it this way: if we're wrong, we can do something else. If they're wrong, we're dead."

Reference (Source) Notes

CHAPTER I: The Credibility Gap

Pages 5–6:
(background on the Corral Canyon fight) Novick, Sheldon, "The Careless Atom," Houghton Mifflin Company, Boston, 1969. Curtis, Richard, and Hogan, Elizabeth, "Perils of the Peaceful Atom," Doubleday and Company, Garden City, N.Y., 1969.
(background, California and earthquakes) "Earthquake," cover story, *Time*, Sept. 1, 1975; Martin Koughan, "Earthquake!" *Reader's Digest*, Nov. 1975 (condensed from *Harper's*, Sept. 1975).

Pages 6–7:
(Bodega Head controversy) Novick, "The Careless Atom"; Curtis and Hogan, "Perils of the Peaceful Atom"; Marine, Gene, "Taking the Initiative in California," *Ramparts*, Summer 1974; Pesonen, David E., "The Ticklish Statistics," *Nation*, Oct. 18, 1965.

CHAPTER II: Legislative Warfare

Page 9:
(California initiative qualifies) statement by March Fong Eu, California secretary of state, May 6, 1975.
(specifics of initiative) Title 7.8., Land Use Nuclear Power Liability and Safeguards Act.

Page 10:
(tests of emergency cooling systems; cost for a full-scale test) "Nuclear Power and the Environment," seventh in a series of fact sheets presented by the Delaware Valley Section of the American Nuclear Society, 1972.
(amount of plutonium produced) phone conversation, Jan. 23, 1976, with Dr. Donald Kuhn, Energy Research and Development Agency's (ERDA) Office of Nuclear Fuel Cycle and Production, Germantown, Md.

Page 13:
(political campaign reform) People's Lobby brochure, 1974, reprinting quotes from various sources.

Page 14:
(Robert Finch quote) *San Diego Union,* May 9, 1975, "Reactors Have Good Record."
(Sigmund Arywitz quote) letter to "All Unions and Councils," Jan. 22, 1975, by Arywitz.
(Pat Brown quote) press release, Citizens for Jobs and Energy, May 6, 1975.

Page 16:
(antinuclear campaigns) Western Bloc newsletter, Jan. 12, 1976.

CHAPTER III: Nuclear Plants

Page 18:
(description, facts about Peach Bottom plant) "Final Environmental Statement Related to Operation of Peach Bottom Units 2 and 3, Philadelphia Electric Company," Docket Nos. 50–277 and 50–278, AEC, 1973; "Peach Bottom Nos. 2 and 3," Philadelphia Electric Company, 1973; Olson, McKinley C., "The Hot River Valley," *Nation,* Aug. 3, 1974.

Page 19:
(radioactivity produced by nuclear plants) Gofman, John W., and Tamplin, Arthur R., "Poisoned Power," Rodale Press, Emmaus, Pa., 1971; Ford, Daniel F., and Kendall, Henry W., "The Nuclear Power Issue: An Overview," *Union of Concerned Scientists,* Jan. 1974.
(Kendall quote) Ford and Kendall, "An Evaluation of Nuclear Reactor Safety," *Union of Concerned Scientists,* March 1972.

Pages 20–21:
(atom bombs and plants) Loftness, Robert L., "Nuclear Power Plants," D. Van Nostrand Co., Princeton, N.J., 1964; Lapp, Ralph, "Atoms and People," Harper & Brothers, New York, 1956; Davis, George E., "Radiation and Life," Iowa State University Press, Ames, 1967; Gofman and Tamplin, "Poisoned Power."
(radiation effects) Frigerio, Norman C., "Your Body and Radiation," AEC, revised, 1967.

Page 21:
(Fermi meltdown) Curtis and Hogan, "Perils of the Peaceful Atom"; Fuller, John G., "We Almost Lost Detroit," Reader's Digest Press, 1975; Sloan, Allan, "Fermi Nuclear Plant Will Close," *Detroit Free Press,* Nov. 31, 1972; Cohn, Victor, "Troubled Michigan Nuclear Plant to Close," *Washington Post,* Nov. 30, 1972.

Page 22:
(accident studies) WASH-740, "Theoretical Possibilities and Consequences of Major Accidents in Large Nuclear Plants," AEC, March 1957—known as the Brookhaven report; WASH-740 update, AEC, 1964–65; Fuller, "We Almost Lost Detroit."
(WASH-740 update disclosed) news release, Union of Concerned

Scientists, Aug. 14, 1973. Also WASH-1400, "An Assessment of Accident Risks in U.S. Commercial Nuclear Power Plants," Draft AEC, Aug. 21, 1974, known as the Rasmussen report.

Pages 23–24:
(responses to Rasmussen report) Hines, William, "Study for AEC Minimizes Perils of Nuclear Plant Mishap," *Chicago Sun-Times,* Aug. 21, 1974; Lyon, Richard D., "AEC Study Finds Hazards of Reactors Very Slight," *New York Times,* Aug. 21, 1974; "Questions Remain on Nuclear Safety," *Milwaukee Journal* editorial, Aug. 26, 1974; Rasmussen, Norman C., "The Safety Study and Its Feedback," *Bulletin of the Atomic Scientists,* Sept. 1975; von Hippel, Frank, "A Perspective on the Debate," *Bulletin of the Atomic Scientists,* Sept. 1975; Hall, Tom, "Dialog: David Comey," *Chicago Tribune,* May 11, 1975; Kendall, Henry, "Nuclear Power Risks: A Review of the Report of the American Physical Society's Study Group on Light Water Reactor Safety," *Union of Concerned Scientists,* June 18, 1975; Abbots, John, "NRC Peddles Reactor Safety Study: Final Report Buries Criticisms," *Critical Mass,* Dec. 1975; von Hippel, "Nuclear Reactor Safety," *Bulletin of the Atomic Scientists,* Jan. 1976.

Pages 24 ff.:
(nuclear reactors and how they work) "Nuclear Power Plants," AEC, 1973 (revised); Loftness, "Nuclear Power Plants."

Page 28:
(thermal pollution) "Thermal Effects and U.S. Nuclear Power Stations," AEC, Aug. 1971; Weil, George L., "Nuclear Energy: Promises, Promises," George Weil, April 1973 (revised).

Page 29:
(Ayers quote) Hall, Tom, "Dialog: Thomas G. Ayers," *Chicago Tribune,* June 8, 1975.
(Alfvén quote) Alfvén, Hannes, "Energy and Environment," *Bulletin of the Atomic Scientists,* May 1972.

Pages 29 ff.:
(emergency core cooling systems) Ford and Kendall, "Evaluation of Nuclear Reactor Safety"; "An Assessment of the Emergency Core Cooling Systems Rulemaking Hearing," *Union of Concerned Scientists,* April 1973; Finlayson, Fred C., "A View from the Outside," *Bulletin of the Atomic Scientists,* Sept. 1975; "Dialog: David Comey," *Chicago Tribune.*

Page 31:
(Kendall) Kendall, "Nuclear Power Risks."

Page 32:
(American Nuclear Society quotes) "Nuclear Power and the Environment," San Diego Section, American Nuclear Society, 1972.
(Aerojet quote) Burnham, David, "AEC Documents Show a 10-Year Effort by Agency to Conceal Studies on Safety Peril Posed by Reactors," *New York Times,* Nov. 10, 1974.

Pages 32–33:
(Browns Ferry fire) press release, Nuclear Regulatory Commission (NRC), April 3, 1975; Comey, David D., "How We Almost Lost Alabama," *Chicago Tribune,* Aug. 31, 1975; Comey, "The Incident at Browns Ferry," *Not Man Apart,* mid-Sept. 1975; "Candle Blows Out Browns Ferry 1, 2, and—Almost—3," *Not Man Apart,* April 1975; "Atomic Plant Fire May Force Changes," *Baltimore Sun,* Sept. 22, 1975; Burnham, David, "Inquiry on Fire at Biggest Nuclear Plant Finds Prevention Program 'Was Essentially Zero,' " *New York Times,* Feb. 29, 1976.

Page 34:
(Lapp quote) Lapp, Ralph, "Nuclear Salvation or Nuclear Folly?" *New York Times Magazine,* Feb. 10, 1974.
(pronuclear views) Bethe, Hans A., "The Necessity of Fission Power," *Scientific American,* Jan. 1976; Teller, Edward, "A Faustian Bargain?" *Development Forum,* Dec. 1974; Lapp, "The Safety of Nuclear Fission Power," debate on Nuclear Power, Madison, Wis., Jan. 17, 1975; McElheny, Victor K., "Hans Bethe Urges Atom Power Drive," *New York Times,* Dec. 14, 1974.
(Alfvén quote) Alfvén, *Bulletin of the Atomic Scientists,* May 1972.

Pages 35–36:
(GE engineers resign) Burnham, David, "3 Engineers Quit G.E. Reactor Division and Volunteer in Antinuclear Movement," *New York Times,* Feb. 3, 1976; "Three Quit Jobs at GE, Cite 'Nuclear Risk,' " *Chicago Tribune,* Feb. 3, 1976; "3 Quit A-Energy Jobs over 'High Risk,' " *Chicago Sun-Times,* Feb. 3, 1976 (from Larry Pryor story in *Los Angeles Times*); "Reactors Less Regulated Than Toasters, 3 Charge," *Chicago Sun-Times,* Feb. 19, 1976; Bukro, Casey, "Phase Out Atom Plants, Ex-Edison Aide Warns," *Chicago Tribune,* Feb. 13, 1976.

Page 36:
(worst accident) Curtis and Hogan, "Perils of the Peaceful Atom"; Loftness, "Nuclear Power Plants"; Novick, "The Careless Atom"; Fuller, "We Almost Lost Detroit."
(mishaps at nuclear plants) Burnham, David, "AEC Penalizes Few Nuclear Facilities Despite Thousands of Safety Violations," *New York Times,* Aug. 25, 1974.

Page 37:
(Comey's assessment) *BPI Newsletter,* Spring 1975.
(Hocevar quotes) Burnham, "Power Reactors Face Safety Test," *New York Times,* Sept. 22, 1974; Scharff, Ned, "As Area's Nuclear Power Plants Grow So Do Fears of Catastrophe," *Washington Star-News,* Feb. 3, 1975.
(Commonwealth Edison plants) *BPI Newsletter,* Spring 1975.
(reactor shutdowns) "Nuclear Regulatory Commission Action Requiring Safety Inspections Which Resulted in Shutdown of Certain Nuclear Power Plants," joint hearing before the Joint Committee on Atomic Energy and the Committee on Government Operations, Feb.

5, 1975; Stockton, William, "Reactor Pipe Cracks Baffle Experts on Nuclear Plants," *Washington Post*, April 13, 1975.

Page 38:
(control-room operator quote) "Nuclear Power Plants," ERDA, revised, 1973.
(Zion plant) *BPI Newsletter*, Spring 1975; Gershen, Martin, "15,000 Gal. of Radioactive Water Leaks in Zion Plant," *Chicago Sun-Times*, June 6, 1975; letter from NRC to David Comey, June 25, 1975, showing 311 Commonwealth Edison violations of NRC regulations Jan. 1974–April 1975; Bukro, Casey, "U.S. Warns Edison on Nuclear Safety," *Chicago Tribune*, Aug. 15, 1975; "9-Minute Radioactive Leak at Zion," *Chicago Sun-Times*, Aug. 22, 1975; Constantine, Peggy, "Edison Nuclear Cooling Problems Disputed," *Chicago Sun-Times*, Sept. 3, 1975.
(Vermont Yankee shutdowns) Kifner, John, "Shutdown of a Nuclear Plant, 17th in 19 Months, Spurs U.S. Debate," *New York Times*, March 31, 1974.
(Consolidated Edison troubles) Fialka, John, "Everything Went Wrong," *Washington Star-News*, April 24, 1974.
(Oyster Creek) Olson, "Hot River Valley," *Nation*, Aug. 3, 1974.
(Millstone reactor leak) "Evacuate Workers After A-Plant Spill," *Chicago Daily News*, March 28, 1975; "Millstone Around the Neck," *Not Man Apart*, May 1975.

Page 39:
(Metropolitan Edison leaks) Baer, John, "3 'Minor' Leaks of Radioactive Gas 'Not Told' to State," *Harrisburg Evening News*, Oct. 17, 1974; "N-Plant Shut Down Due to Leak of Gas," *Lancaster* (Pa.) *Intelligencer*, Oct. 24, 1974; "Radioactive Gases Released at Nuclear Facility, Met Ed Admits," *Harrisburg Patriot*, Jan. 28, 1975.
(American Nuclear Society quote) "Nuclear Power and the Environment," San Diego Section, American Nuclear Society, 1972.
(North Anna River reactors and earthquake fault) Willard, Hal, "A Geologic Fault Bedevils Reactor," *Washington Post*, Sept. 27, 1973; "Sweet and Sour Virginia," *Not Man Apart*, May 1975; Rosenthal, Bruce, "Va. Utility Guilty of Cover-up," *Critical Mass*, July 1975; Willard, Hal, "Vepco Fined $60,000 for A-Plant Fault," *Washington Post*, Sept. 12, 1975; "Virginia Electric Power Fined $60,000 by NRC," *Not Man Apart*, Nov. 1975.

Page 40:
(amount of money invested in nuclear power) Simpson, John W., director of Westinghouse's reactor division, "Nuclear Energy and the Future," paid position paper, *Fortune*, Feb. 1975.

CHAPTER IV: The Controversial Roots

Pages 41 ff.:
(based on my visit to Grand Junction, interviews there, plus background—unless otherwise indicated—from the following) "Radiation Exposure of Uranium Miners," vols. I–II, Joint Committee on

Atomic Energy, 1967; Metzger, H. Peter, "The Atomic Establishment," Simon and Schuster, New York, 1972; Novick, "The Careless Atom."

Page 48:
(Saccomanno's work) Saccomanno, Geno, et al., "Concentration of Carcinoma or Atypical Cells in Sputum," U.S. Department of Health, Education, and Welfare, from *Acta-Cytologia,* vol. 7, no. 5, Sept.-Oct. 1963.

Page 50:
(uranium tailings in water—discussed in books above, but also) "Water Near Uranium Mines Held Risky," *Chicago Sun-Times,* Aug. 19, 1975, from Stuart Auerback's story in the *Washington Post;* "NRDC Requests EIS on Uranium Mills," *Not Man Apart,* mid-July 1975.

Pages 50 ff.:
(background, history of nuclear power) Lapp, "Atoms and People"; Woodbury, David O., "Atoms for Peace," Dodd, Mead and Co., New York, 1965; Martin, Charles Noël, "The Atom: Friend or Foe," Franklin Watts, New York, 1962; Curtis and Hogan, "Perils of the Peaceful Atom"; Novick, "The Careless Atom."

Page 54:
(Yount quote) "The Careless Atom."

Pages 55–56:
(Kepford testimony) Sept. 24, 1975.
(Denenberg quote, estimate) "A Citizen's Bill of Rights on Nuclear Power," State of Pennsylvania Insurance Department, Sept. 1973; Denenberg's testimony Nov. 7, 1973, before the AEC Safety and Licensing Board hearing for the Three Mile Island Nuclear Power Plant.

Pages 57 ff.:
(background material—unless otherwise indicated—on the Fermi plant) Novick, "The Careless Atom"; Curtis and Hogan, "Perils of the Peaceful Atom"; Fuller, "We Almost Lost Detroit."

Page 58:
(doubts about Fermi) "Safety Doubts May Delay Detroit Reactor, AEC Official Indicates," *Wall Street Journal,* July 20, 1956.

Page 60:
(Hans Bethe testimony) "Physicist Tells AEC Reactor Near Detroit Is Safe Enough to Build," *Wall Street Journal,* Jan. 9, 1957.

Page 61:
(decision blocking Fermi) Finney, John W., "Court Balks AEC on Construction of Power Station," *New York Times,* June 11, 1960.
(Supreme Court upholds AEC) Ottenberg, Miriam, "AEC Is Upheld by High Court on A-Plants," *Washington Star,* June 12, 1961.

Page 62:
(EBR-1 and SL-1 accidents) Loftness, "Nuclear Power Plants."

Page 63:
(cause of Fermi accident) "Eight-Inch Piece of Sheet Metal Triggered Accident in Atomic Plant," *National Observer*, Oct. 5, 1966.

Page 64:
(Fermi closing; stories about Fermi's history) "Fermi 1 Nuclear Plant to Close Near Monroe," *Toledo Blade*, Nov. 29, 1972; "Troubled Michigan Nuclear Plant to Close," *Washington Post*, Nov. 30, 1972; "Fermi Nuclear Plant Will Close," *Detroit Free Press*, Nov. 30, 1972.
(liquid sodium at Fermi plant) Goldberg, Deborah, "Disposal of Radioactive Sodium Puzzles Michigan Power Plant," *Ogdensburg* (N.Y.) *Journal*.

Page 65:
(proposed Ravenswood plant) Novick, "The Careless Atom"; Curtis and Hogan, "Perils of the Peaceful Atom."
(Gofman quote) Gofman and Tamplin, "Poisoned Power."

Pages 66–67:
(fallout controversy; effects of fallout) Lapp, "Atoms and People"; Davis, "Radiation and Life"; Martin, "The Atom: Friend or Foe"; Novick, "Perils of the Peaceful Atom"; Metzger, "The Atomic Establishment"; Curtis and Hogan, "Perils of the Peaceful Atom."

Pages 67 ff.:
(Tamplin-Gofman controversy; their findings) Gofman and Tamplin, "Epidemiologic Studies of Carcinogenesis by Ionizing Radiation," *Proceedings of the Sixth Berkeley Symposium on Mathematical Statistics and Probability*, University of California Press, June-July 1970; Gofman and Tamplin, "Poisoned Power."

Page 68:
(concern about atomic workers) Ibser, H. W., "The Nuclear Energy Game: Genetic Roulette," *Progressive*, Jan. 1976.
(nuclear industry response to Gofman-Tamplin) "Nuclear Power and the Environment," American Nuclear Society, 1974.

Page 69:
(David Levin response to Gofman-Tamplin study) *Proceedings of the Sixth Berkeley Symposium on Mathematical Statistics and Probability*, University of California Press, 1970.

Page 70:
(Gofman quote) Gofman, "Alice in Blunderland," address, Nuclear Energy Forum, San Luis Obispo, Calif., Oct. 17, 1975.

Page 71:
(Lilienthal quote) "Fact Sheet," Senator Mike Gravel's office, 1974.

Pages 73 ff.:
(interviews with Comey and Cherry in May-June 1975, backed up by examination of their press and magazine clips)
(*Nuclear Journal* quote about Comey) "David Comey Combines Dedication with a Thirst for Knowledge," *Nuclear Journal*, April 1973.

Page 74:
(Comey EPA award) *BPI Newsletter*, Spring 1975.

Page 77:
(summation of Palisades case, and effects) "Nuclear Power Plants: Do They Work?" *BPI Annual Report*, 1973.

Pages 78 ff.:
(Ford and Kendall, Union of Concerned Scientists, ECCS hearings) Zeman, Tom, "Lying Doesn't Make It Safe," *Ramparts*, Summer 1974; Ford and Kendall, "The Nuclear Power Issue: An Overview," *Union of Concerned Scientists*, Jan. 1974; Ford and Kendall, "An Evaluation of Nuclear Reactor Safety," vols. 1–2, *Union of Concerned Scientists*, March, Oct. 1972; Ford and Kendall, "An Assessment of the Emergency Core Cooling Systems Rulemaking Hearing," *Union of Concerned Scientists*, April 1973; Gofman, John, "Time for a Moratorium," from "The Case for a Nuclear Moratorium," Environmental Action Foundation, 2d ed., 1st printing, May 1974.

Page 82:
(Calvert Cliffs case) Weil, "Nuclear Energy: Promises, Promises," published by George Weil, 3d printing, revised, April 1973; *Calvert Cliffs' Coordinating Committee* v. *AEC*, 449 F.2d 1109, 1128, Circuit Court of Appeals, Washington, D.C., 1971.

Page 83:
(suit against AEC) *Nader and Friends of the Earth* v. *Dixy Lee Ray et al., AEC*, May 31, 1973, in U.S. District Court, Washington, D.C.; judge dismissed the suit June 28, 1973, in effect, turning it back to the AEC; case was appealed to no avail.

Page 84:
(Comey's economics) Comey, "Nuclear Power Plant Reliability," report to the Federal Energy Administration, Sept. 10, 1974.
(AEC safety cover-up) Burnham, David, "AEC Documents Show a 10-Year Effort by Agency to Conceal Studies on Safety Peril Posed by Reactors," *New York Times*, Nov. 10, 1974.
(AEC abolished) "AEC Abolished as Ford Sets Up a New Agency," *Harrisburg Patriot*, Jan. 11, 1975.

Pages 85–86:
(Bailey case, evacuation issue) "Court Blocks Plan for Nuclear Plant Near Lake Michigan," *Wall Street Journal,* April 2, 1975; King, Seth S., "Court Bars Nuclear Reactor in Densely Settled Area," *New York Times,* April 3, 1975; Bukro, Casey, "No More Atomic Plants Likely on Lake, Official Says," *Chicago Tribune,* April 10, 1975; Sirico, Lou, "Citizens Demand Improved Evacuation Procedures," *Critical Mass,* Aug. 1975; Lanouette, William J., "Evacuation After a Nuclear Accident: Can It Be Done?" *National Observer,* July 12, 1975; Comey, David, "Do No Go Gentle into That Radiation Zone," *Bulletin of The Atomic Scientists,* Nov. 1975.

CHAPTER V: Radioactive Implications

Page 87:
(dismiss low-level radiation dangers) Grey, Jerry, "Low Level Radiation," *Atomic Industrial Forum,* April 1974; "Consultant Belittles Nuclear Plant Dangers," AP story, *Wisconsin State Journal,* Sept. 11, 1974; "Nuclear Power and the Environment," American Nuclear Society, March 1974.

Page 88:
(McCormack—spokesman for the "acceptable risk" theory; dismissing critics on low-level radiation issue) McCormack, Mike, "Energy, Environment and the Economy," *Congressional Record—House,* Dec. 4, 1974; address before the National Security Industrial Association, April 9, 1975.
(Martell: first quote) Boulder Colo., interview, June 1974—repeated in dozens of phone calls through Jan. 1976.

Page 89:
(Martell: second quote) Jan. 1976; writings: Basic Considerations in the Assessment of the Cancer Risks and Standards for Internal Alpha Emitters," statement, public hearings on plutonium standards, Environmental Protection Agency (EPA), Denver, Colo., Jan. 10, 1975; "Tobacco Radioactivity and Cancer in Smokers," *American Scientist,* July-Aug. 1975.
(Gofman) Gofman, John W., "The Cancer Hazard from Inhaled Plutonium," Committee for Nuclear Responsibility, May 14, 1975; "Estimated Production of Human Lung Cancers by Plutonium from Worldwide Fallout," Committee for Nuclear Responsibility, July 10, 1975.

Page 90:
(McCormack quote) McCormack, *Congressional Record—House,* Dec. 4, 1974.

Page 91:
(americium and curium threat) Martell, Edward, and Poet, S. E., "Plutonium-239 and Americium-241 Contamination in the Denver Area," *Health Physics,* vol. 23, Oct. 1972; "Recent Animal Studies on the Deposition, Retention and Translocation of Plutonium and

Other Transuranic Compounds," Bair, W. J., Battelle, Pacific North-west Laboratories Paper, written under contract (E-45–1–1830) for ERDA; Martell, "Actinides in the Environment and Their Uptake by Man," Atmospheric Quality and Modification Division, National Center for Atmospheric Research, May 1975.

(amount of plutonium produced) phone interview, Jan. 23, 1976, with Dr. Donald Kuhn, ERDA's Office of Nuclear Fuel Cycle and Production, Germantown, Md.

(Gofman quote) Gofman, "The Cancer Hazard from Inhaled Plutonium."

(Geesaman) Geesaman, Donald P., "An Analysis of the Carcinogenic Risk from an Insoluble Alpha-Emitting Aerosol Deposited in Deep Respiratory Tissue," reports of the Lawrence Livermore Laborratory for the AEC, Feb. 9, Oct. 9, 1968; "For Science Students: A Quiet Polemic," presented by Geesaman at a meeting of the Minnesota Science Teachers' Association, Oct. 18, 1974.

Page 92:
(Weinberg quote) Weinberg, Alvin M., "Social Institutions and Nuclear Energy," *Science,* July 7, 1972.

(different kinds of radiation) Gofman and Tamplin, "Poisoned Power"; Davis, "Radiation and Life"; Frigerio, "Your Body and Radiation"; "Nuclear Power and the Environment," American Nuclear Society.

Page 93:
(Lapp estimate) Grey, "Low-Level Radiation," and McCormack, "Energy, Environment and the Economy."

Page 94:
(radiation dosage) Frigerio, "Your Body and Radiation"; Gofman and Tamplin, "Poisoned Power."

(background radiation) "Nuclear Power and the Environment," American Nuclear Society.

Page 95:
(no amount of radiation safe) Muller, H. J., "Radiation Damage to the Genetic Material," Science in Progress Series, Yale University Press, 1950; Pauling, Linus, "Genetic and Somatic Effects of High-Energy Radiation," *Bulletin of the Atomic Scientists,* Sept. 1970.

(Pauling estimate) same source as above.

(cancer statistics) " '76 Cancer Facts and Figures," American Cancer Society, 1975; "Fact Book 1975," National Cancer Institute, revised, Jan. 1975.

Page 96:
(radiation effects) Frigerio, "Your Body and Radiation."

(Lederberg quotes) from Gofman and Tamplin's "Poisoned Power," pp. 76, 85.

Page 97:
(Martell) phone interview, Jan. 1976.

(as "low as practicable") "Nuclear Power and the Environment,"
American Nuclear Society.

Page 98:
(radioactivity and the human food chain) Gofman and Tamplin,
"Poisoned Power"; Novick, "The Careless Atom"; Curtis and Hogan,
"Perils of the Peaceful Atom"; Martell, "Actinides in the Environ-
ment and Their Uptake by Man."
(Gofman's calculations about radioactivity from nuclear plants)
Gofman, "The Case Against Nuclear Power," *Catalyst*, vol. II, no.
3, 1971.

Page 99:
(McCormack estimate) McCormack, "Energy, Environment and the
Economy."

Pages 99 ff.:
(Sternglass) Sternglass, E. J., "Significance of Radiation Monitoring
Results for the Shippingport Nuclear Reactor," Department of
Radiology, University of Pittsburgh, Jan. 21, 1973; Sternglass, "Ra-
dioactive Waste Discharges from the Shippingport Nuclear Power
Station and Changes in Cancer Mortality," University of Pittsburgh,
May 8, 1973; Griffiths, Joel, "America's Biggest Nuclear Power
Scandal!" *Saga*, Oct. 1973.
(refuting Sternglass) Grey, "Low-Level Radiation."

Page 101:
(Pennsylvania report on Sternglass's allegations) press release, June
26, 1974, Division of Comprehensive Health Planning, Department of
Health, Commonwealth of Pennsylvania.

Pages 102 ff.:
(based on in-person interviews with Martell in Boulder in June 1975,
Chicago, Nov. 1975, numerous phone interviews, plus following background sources) Gofman and Tamplin, "Poisoned Power"; David,
"Radiation and Life"; Martell, "Basic Considerations in the Assessment of the Cancer Risks and Standards of Internal Alpha Emitters"
and "Tobacco Radioactivity and Cancer in Smokers."

Pages 106–107:
(the way alpha radiation works; government standards, etc.) two
Martell articles above, plus numerous phone calls, etc.; Gofman,
"The Cancer Hazard from Inhaled Plutonium"; Cochran, Thomas B.,
Tamplin, Arthur, and Speth, J. Gustave, "The Plutonium Decision:
A Report on the Risks of Plutonium Recycle," Natural Resources De-
fense Council, Sept. 1974.

Page 107:
("pronuclear" view) Bair, W. J., and Thomas, J. M., "Prediction
of the Health Effects of Inhaled Transuranium Elements from Experimental Data," Battelle, Pacific Northwest Laboratories, based on
work for ERDA, contract (E-45-1-1830).

(Simpson quote) Simpson, John, "Nuclear Energy and the Future," *Fortune*, May 1975.

Page 108:
(Martell and fallout) Metzger, "The Atomic Establishment."

Pages 109 ff.:
(Rocky Flats plant, fire, etc.) Metzger, "The Atomic Establishment"; Martell, E., Goldan, P. A., Kraushaar, J. J., Shea, D. W., and Williams, R. H., "Fire Damage," *Environment*, vol. 12, no. 4, May 1970 —with "AEC Statement" at end of article; interviews at Rocky Flats, June 1975.

Page 112:
(Rocky Flats monitoring for radioactivity) Martell and Poet, "Plutonium-239 and Americium-241 Contamination in the Denver Area." (Martell quote) from *Health Physics* article (above) written in 1971, published in 1972.
(Rocky Flats report) "Annual Environmental Monitoring Report, Rocky Flats Plant," Dow Chemical U.S.A., for ERDA, April 30, 1975.

Page 113:
(Martell 1972 quote) Martell, "Safety at Rocky Flats," *Science*, Oct. 27, 1972.
(Martell 1975 quote) Martell, "Actinides in the Environment and Their Uptake by Man."

Pages 113 ff.:
(distilled from these sources) Martell, "Radioactivity of Tobacco Trichomes and Insoluable Cigarette Smoke Particles," *Nature*, vol. 249, no. 5454, May 17, 1974; "Radioactive Smoke Particles; New Link to Lung Cancer?" National Center for Atmospheric Research, May 17, 1974; "Aerosol Particles on Tobacco Trichomes," *Nature*, vol. 270, July 12, 1974; Martell, "Basic Considerations in the Assessment of the Cancer Risks and Standards for Internal Alpha Emitters"; "Actinides in the Environment and Their Uptake on Man"; "Tobacco Radioactivity and Cancer in Smokers."

Pages 117 ff.:
(Martell 1975 statement) *American Scientist*, July-Aug. 1975.
(interviews with Martell; his writings) "Basic Considerations in the Assessment of the Cancer Risks and Standards for Internal Alpha Emitters"; "Actinides in the Environment and their Uptake by Man"; "Tobacco Radioactivity and Cancer in Smokers."

Page 121:
(dust and urban pollution) Draeger, Harlan, "Chicago's Soot Villain —Heavy Auto Traffic," *Chicago Daily News*, Aug. 20, 1975.

Page 122:
(letter from McCormack) *Not Man Apart*, mid-Aug. 1975.

(McCormack: no one's been hurt) address at Notre Dame University, April 14, 1975; reprinted, *Congressional Record—House,* April 17, 1975.
(McCormack estimation of plutonium fallout) my interview with him, May 1975; *National Journal Reports,* April 5, 1975, p. 512.

Pages 122 ff.:
(Gofman and plutonium—my interview, June 1975, numerous subsequent phone interviews, and mainly these sources) Gofman's reports for Committee for Nuclear Responsibility: "The Cancer Hazard from Inhaled Plutonium," May 14, 1975; "Estimated Production of Human Lung Cancers by Plutonium from Worldwide Fallout," July 10, 1975.

Page 127:
(Los Alamos study) Hempelman, L. H., Langham, W. H., Richmond, C. R., and Voelz, G. L., "Manhattan Project Plutonium Workers: A Twenty-seven Year Follow-up Study of Selected Cases," *Health Physics,* vol. 25, Nov. 1973.

Pages 127 ff.:
(Martell assessment) Martell's "Basic Considerations in the Assessment of the Cancer Risks and Standards for Internal Alpha Emitters" (his assessment of Los Alamos workers); radiation effects of plutonium—interviews plus "Actinides in the Environment and Their Uptake on Man"; "Tobacco Radioactivity and Cancer in Smokers."

Pages 129 ff.:
(Rocky Flats health hearing) Brimberg, Judith, "Dow Workers' Radiation Exposure Called High," *Denver Post,* Aug. 14, 1970.
(interviews with Brown and Nicks) Rocky Flats plant, June 1975.

Pages 131 ff.:
(Karen Silkwood case: background material, unless otherwise specified) Burnham, David, "Death of Plutonium Worker Questioned by Union Official," *New York Times,* Nov. 19, 1974; Burnham, David, "Plutonium Plant Under Security," *New York Times,* Nov. 20, 1974; McCarthy, Tom, "McGee Assails Claims," *Daily Oklahoman,* Nov. 20, 1974; "Probe Is Sought in Traffic Death," *Tulsa Daily World,* Nov. 20, 1974; Hearn, Gary, "Union Complained to AEC Weeks Before Contamination," *Oklahoma City Times,* Nov. 21, 1974; McCarthy, Tom, "Justice Officials to Probe Nuclear Workers' Case," *Daily Oklahoman,* Nov. 21, 1974; "Possible Plutonium Theft at Nuclear Plant Probed," *Des Moines Register* (from the *Washington Post*), Dec. 9, 1974; McCarthy, Tom, "Autopsy Shows Atomic Worker Under Drug's Influence," *Daily Oklahoman,* Nov. 22, 1974; Kohn, Howard, "The Nuclear Industry's Terrible Power and How it Silenced Karen Silkwood," *Rolling Stone,* March 27, 1975; Phillips, B. J., "The Case of Karen Silkwood," *Ms.,* 1975; "Plutonium Worker Advocated Safety," *Critical Mass,* April 1975.

Page 132:
(Younghein quote) "Boomer, Sooner," Younghein's newsletter, 1975.

Page 135:
(congressional investigation of Silkwood death) *Critical Mass*, Dec. 1975.
(*New York Times* story of missing plutonium) Burnham, David, "Thousands of Pounds of Materials Used in Nuclear Bombs Are Unaccounted for," *New York Times*, Dec. 29, 1974.
(allegations about Kerr-McGee substantiated) "AEC Investigation Completed of Contamination Incident at Oklahoma Nuclear Plant," AEC press release, Jan. 6, 1975; "AEC Investigations Completed on Allegations Concerning Working Conditions and Quality Assurance at Kerr-McGee Plutonium Facility in Oklahoma," AEC press release, Jan 7, 1975; AEC's Investigation Report Nos. 070–925/74–05; 070–1193/74–10 and 040–7308/74–03, reviewed Dec. 13, 1974; and AEC investigation report No. 74–09, reviewed Dec. 16, 1974; Burnham, David, "AEC Finds Evidence Supporting Charges of Health Hazards at Plutonium Processing Plant in Oklahoma," Jan 8, 1975.
(Kerr-McGee plant closing) "Kerr-McGee Plant to Shut Down," *Not Man Apart*, mid-Nov. 1975; "Will We Ever Know What Happened at Kerr-McGee?" *Not Man Apart*, mid-Jan. 1976.
(phone interviews with Oklahoma City reporters, Jack Taylor, *Oklahoma Times;* Judy Fossett, *Oklahoma Times;* Alan Bromley, *Daily Oklahoman;* Ilene Younghein, June 1975)

Page 136:
(*Daily Oklahoman* quotes) "Anti-Nuclear Delusions," editorial, *Daily Oklahoman*, June 6, 1975; "Banning the Future," editorial, *Daily Oklahoman*, Jan. 9, 1975.
(Mazzocchi quote) phone interview, June 1975.

Page 137:
(Martell quote) Martell, Edward, "Neglected Aspects of the Health Effects of Low Levels of Ionizing Radiation on Man," statement submitted to the National Research Council Committee on Nuclear and Alternative Energy Systems, Feb. 1976.
(Gio Batta Gori comments) Phone interview, March 31, 1976.

Pages 138 ff.:
(phone interviews) Barr interview, Jan. 23, 1976; Bair interview, Jan. 23, 1976; Raabe interview, Jan. 29, 1976.

Page 140:
(reading materials supplied by Bair) Sanders, C. L., and Adee, R. R., "Phagocytosis of Inhaled Plutonium Oxide-239 Particles by Pulmonary Macrophages," *Science*, vol. 162, Nov, 22, 1968: Bair, W. J., "Inhalation of Radionuclides and Carcinogenesis," Battelle Northwest Laboratories, for AEC Symposium Series on *Inhalation Carcinogenesis*, Springfield, Va., 1970; Park, J. F., and Sanders, C. L.,

"Pulmonary Distribution of Alpha Dose from Plutonium Dioxide and Induction of Neoplasia in Rats and Dogs," *Inhaled Particles III*, vol. 1, Unwin Brothers, Surrey, England, 1971; Sanders, C. L., and Jackson, T. A., "Induction of Mesotheliomas and Sarcomas from 'Hot Spots' of Plutonium Dioxide Activity," *Health Physics*, vol. 22, June 1972; Sanders, Charles L., and Dagle, G. E., "Studies of Pulmonary Carcinogenesis in Rodents Following Inhalation of Transuranic Compounds," edited by Karbe, E., and Park, J. F., in "Experimental Lung Cancer, Carcinogenesis and Bioassays," Springer-Verlag, Berlin, Heidelberg, New York, 1974; Craig, D. K., Mahlum, D. D., and Klepper, E. L., "The Relative Quantity of Airborne Plutonium Deposited in the Respiratory Tract and on the Skin of Rats," *Health Physics*, vol. 27, Nov. 1974; Bair, W. J., and Thompson, R. C., "Plutonium: Biomedical Research," *Science*, vol. 183, Feb. 22, 1974; Bair, W. J., "Considerations in Assessing the Potential Harm to Populations Exposed to Low Levels of Plutonium in Air," from "Population Dose Evaluation and Standards for Man and His Environment," International Atomic Energy Agency, Vienna, 1974; Sanders, C. L., Jr., Thompson, R. C., and Bair, W. J., "Lung Cancer: Dose Response Studies with Radionuclides," work performed by Battelle Northwest Laboratories, AEC contract (E-45-1-1830), delivered at University of California San Francisco Medical Center, Session III, "Respiratory Carcinogenesis."

Page 141:
(cancer threat) Brody, Jane E., "Experts Baffled by Rise in Cancer," *New York Times*, Nov. 28, 1975; Hines, William, "Dawning of Age of Cancer Is Near, Researcher Says," *Chicago Sun-Times*, Jan. 13, 1976.

CHAPTER VI: The "Hot" Fuel Cycle

Page 143:
(facts about Hanford) information on display at the Hanford Science Center.

Page 144:
(amount of radioactivity produced by nuclear plant) Gofman and Tamplin, "Poisoned Power"; Ford and Kendall, "The Nuclear Power Issue: An Overview."
(Gofman quote) Spake, Amanda, "South Carolina's Silent Death Factory," *New Times*, 1975.

Page 145:
(Westinghouse's claim for the breeder) Westinghouse Electric Corporation, "Our Only Reasonable Alternative: The Case for the Liquid Metal Fast Breeder Reactor," 1975.

Page 146:
("rapid decline" quote) Acting Federal Energy Administration (FEA) Deputy Administrator John Hill, testifying before Joint

Committee on Atomic Energy, from Bureau of National Affairs, May 8, 1975.
(George White) phone interview, Jan. 14, 1976.

Pages 147 ff.:
(background material for uranium supplies) Getschow, George, "Utilities to Face a Scramble for Uranium as More Nuclear Plants Come on Stream," *Wall Street Journal,* Sept. 1974; "It Works for OPEC; Watch Out for the UPF," *Not Man Apart,* March 1975; Phillips, James G., "Energy Report: Breeder Reactor Continues To Receive Administration Priority, *National Journal Reports,* March 1, 1975; Chernow, Ron, "U.S. Confronts Uranium Crisis," *Philadelphia Inquirer,* May 5, 1975; "Uranium Backfires on Westinghouse," *Business Week,* Aug. 18, 1975; Day, M. C., "Nuclear Energy: A Second Round of Questions," *Bulletin of the Atomic Scientists,* Dec. 1975; "Why Atomic Power Dims Today," *Business Week,* Nov. 17, 1975; "Westinghouse Defaults on Uranium Contracts," *Not Man Apart,* mid-Nov. 1975; "Lapp, Others Warn of Uranium Shortage" and "Can the U.S. Find Nine New Colorado Plateaus?" *Not Man Apart,* Dec. 1975; Ingersoll, Bruce, "Commonwealth Edison Is Taking Up Uranium Prospecting," *Chicago Sun-Times,* Dec. 22, 1975; "U.S. Needs Nine New Colorado Plateaus to Fill Uranium Need," *Not Man Apart,* mid-Jan. 1976; "GAO Warns of Uranium Shortages, *Critical Mass,* Feb. 1976.
(James Duree) phone interview, June 1975.

Pages 149 ff.:
(background for uranium enrichment) Phillips, James G., "Energy Report: Safeguards, Recycling Broaden Nuclear Power Debate," *National Journal Reports,* March 22, 1975; Olson, McKinley C., "Nuclear Fuel: The Hot Shuffle," *Progressive,* April 1975; "Rubber Vendor Goes Nuclear," *Not Man Apart,* May 1975; Nordlinger, Stephen E., "Ford Wants to Let Industry Sell A-Fuels," *Baltimore Sun,* June 27, 1975; Young, Robert, "Ford Asks for Private Atom Plants," *Chicago Tribune,* June 27, 1975; terHorst, Jerald, "Ford Opens New Nuclear Era," column, *Chicago Tribune,* July 13, 1975; Lewis, Paul, "A Nuclear Peace—Profit Motive," *New York Times,* July 20, 1975; "End of a Monopoly?" *Time,* June 30, 1975; "Why Atomic Power Dims Today," *Business Week,* Nov. 17, 1975.

Pages 152 ff.:
(breeder controversy) "Fast Flux Test Facility," Westinghouse Hanford Company, 1975; "Our Only Reasonable Alternative," Westinghouse Electric Corporation, 1975; Cowan, Edward, "AEC Is Criticized on 'Breeder' Plan," *New York Times,* May 6, 1974; Cowan, Edward, "On Risks of Breeder Reactors," *New York Times,* April 23, 1974; "Nuclear Energy Hopes Rest on Brave New World," *Washington Star-News,* April 26, 1974; Novick, Sheldon, "Nuclear Breeders," *Environment,* July-Aug. 1974; Bukro, Casey, "AEC Rejects Bid for Breeder Reactor," *Chicago Tribune,* Nov. 28, 1974; McElheny, Victor K., "AEC Reveals Size of Uranium Stock," *New York Times,* Jan. 5, 1975; "Suddenly the Gas-Cooled Breeder Looks Good," *Business Week,* Feb. 17, 1975; Burnham, David, "Congress Faces Key Deci-

sions on Nuclear Reactors," *New York Times,* March 30, 1975; Phillips, James G., "Energy Report: Breeder Reactors Continues to Receive Administration Priority," *National Journal Reports,* March 3, 1975; Gillette, Robert, "Atomic Energy Will Be a Prime Source of Debate," *New York Times,* Feb. 9, 1975; "Energy; Nuclear Breeder Reactor," *National Journal Reports,* March 1975; Fialka, John, "New Reactors Will Breed Plutonium—and Possibly Trouble," *Washington Star-News,* April 24, 1975; "EPA Puts Breeder on Back Burner," *Not Man Apart,* mid-May 1975; Cubie, Jim, "Congress Weighs Breeder," *Critical Mass,* May 1975; "Nuclear Power: The Case Can Be Overstated," *New York Times* (News in Review), May 5, 1975; "Atomic Energy: Joint Committee Chairman Sees Breeder Almost Doomed After EPA Statement," Bureau of National Affairs, May 8, 1975; "Safety Factors May Delay Use of Breeder Reactors," *Harrisburg Evening News,* May 7, 1975; Young, Robert, "Nuclear Energy Plan 'Breeds' Lobby Alliance," *Chicago Tribune,* June 23, 1975; "The Breeder Reactor: A Premature Decision," *Critical Mass* press release, June 16, 1975; Cubie, Jim, "Rising Congressional Concern Slows Breeder Program," *Critical Mass,* July 1975; Phillips, James G., "Energy Report: Administration Stands Firm on Nuclear Breeder Reactor," *National Journal Reports,* July 5, 1975; "GAO Urges Continued U.S. Research on Breeder Reactor," *Chicago Sun-Times,* July 31, 1975; "Energy Unit to Push Study on Fast-Breeder Reactor," *Chicago Sun-Times,* Jan. 3, 1976 (from Bill Richards story in *Washington Post*).

Pages 154–155:
(plutonium recycle) Ripley, Anthony, "Law Group Calls for Far Stricter Safeguards on Radiation from Plutonium," *New York Times,* Feb. 17, 1974; Cochran, Thomas B., Speth, Gustave J., and Tamplin, Arthur R., "The Plutonium Decision: A Report on the Risks of Plutonium Recycle," Natural Resources Defense Council, Sept. 1974; "Senators Seek Delay in Plutonium Recycling," *Science,* Oct. 11, 1974; "A Rough Month for Plutonium, *Not Man Apart,* mid-March 1975; Phillips, "Energy Report: Safeguards, Recycling Broaden Nuclear Power Debate," *National Journal Reports,* March 22, 1975; "Watching Congress: Plutonium Recycled," *Critical Mass,* April 1975; "NRC Backs Down on Plutonium Recycle Decision," *Critical Mass,* Dec. 1975; "NRDC Sues to Stop Plutonium Recycle," *Not Man Apart,* Jan. 1976; Day, Samuel H., Jr., "Plutonium Recycling," editorial, *Bulletin of the Atomic Scientists,*" Jan. 1976.

Pages 155 ff.:
(fuel reprocessing) "Witnesses Differ on Atom Hazards," *New York Times,* Sept. 22, 1974; Bukro, Casey, "A Setback to Nuclear Planning," *Chicago Tribune,* Aug. 16, 1974; Burnham, David, "Atomic Reactors May Have to Shut," *New York Times,* Jan. 5, 1975; "Nuclear Fuel Services May Never Reopen," *Not Man Apart,* mid-June 1975; Olson, McKinley C., "Nuclear Fuel: The Hot Shuffle," *Progressive,* April 1975; Fialaka, John, "Private Plutonium Business Is Bringing Big Money to the South," *Washington Star-News,* April 23, 1975; Spake, Amanda, "South Carolina's Silent Death Factory," *New Times,* 1975; Spake, Amanda, "A Nuclear Death Factory Is Being Built over Doctors' Objections," *Physicians Management,* June

1975; Ingersoll, Bruce, "Nuclear Dump for the Nation," *Chicago Sun-Times*, Dec. 14, 1975.

Pages 161 ff.:
(radioactive waste) "The Storage of Radioactive Wastes," AEC, 1974; Grey, Jerry, "Managing Nuclear Wastes," *Atomic Industrial Forum*, April 1974; "Radioactive Waste Management at Hanford," Atlantic-Richfield Hanford Company, revised, March 1971; Weinberg, Alvin, "Civilization and Science Symposium," Dec. 1974, p. 121; Alfvén, Hannes, "Energy and the Environment," *Bulletin of the Atomic Scientists*, May 1972; WASH-1520, AEC, 1972—leaking plutonium at Hanford; "Company Criticized by AEC on Leak of Radioactive Waste," *New York Times*, Aug. 5, 1973; "Ultimate Garbage Crisis," *Time*, Aug. 20, 1973; Sterba, James P., "Radiation Traced to Atom Plant in Colorado," *New York Times*, Sept. 27, 1973; McElheny, Victor K., "AEC Aide Finds Waste Kept Safe," *New York Times*, Sept. 27, 1973; Farney, Dennis, "Ominous Problem: What to Do with Radioactive Waste," *Smithsonian Magazine*, April 1974; "Nuclear Power and the Environment," *American Nuclear Society*, March 1974; Dye, Lee, "Deadly Nuclear Garbage Threatens World of Tomorrow," *Los Angeles Times*, May 5, 1974; Olson, McKinley C., "The Hot River Valley," *Nation*, Aug. 3, 1974; McCormack, Mike, "Energy, Environment, and the Economy," *Congressional Record House—House*, Dec. 4, 1974; Simpson, John W., "Managing and Safeguarding Wastes and Fissionable Material," paid position paper, *Fortune*, May 1975; "Ford Cuts Ground from Under Waste Storage Scheme," *Not Man Apart*, May 1975; "Public Has a Right to Know," editorial, *Business Week*, March 10, 1975; "Wastes Disposal Called Threat to Nuclear Energy," *Baltimore Sun*, May 15, 1975; "Hanford Comes Through Again: New Leak in April," *Not Man Apart*, mid-May 1975; "The Ultimate Dump," *Philadelphia Inquirer*, March 9, 1975 (*from* Charles Foley's *London Sunday Observer* story).

Page 165:
(breeder test facility) "Fast Flux Test Facility," Westinghouse Hanford Company, 1975; interview at Hanford in June 1975 with Fred J. Leitz, senior staff scientist, Westinghouse Hanford Company.

Page 166:
("explosive" breeder reactor) Novick, Sheldon, "The Careless Atom"; Teller, Edward, "Fast Reactors: Maybe," *Nuclear News*, Aug. 1967; Cochran, Thomas B., "The Liquid Metal Fast Breeder Reactor: An Environmental and Economic Critique," Johns Hopkins University Press, Baltimore, 1974; review of Cochran's book by Frank von Hippel, *Physics Today*, Dec. 1974; Novick, Sheldon, "Nuclear Breeders," *Environment*, July-Aug. 1974; Fialka, John, "New Reactor Will Breed Plutonium—and Possibly Trouble," *Washington Star-News*, April 24, 1975; Fialka, "Nuclear Power Experts' New Worry: 'The Poof,' " *Washington Star-News*, April 25, 1975.

Pages 166 ff.:
(transportation of nuclear fuel) "How Radioactive Materials Are Shipped," AEC, 1974.

(Gofman quote) Gofman, John, "The Cancer Hazard from Inhaled Plutonium."

(furor over flying plutonium) "Plutonium Warning," editorial, *New York Times*, March 29, 1975; Burnham, David, "Plutonium Shipment to City Approved," *New York Times*, April 1, 1975; Burnham, "Two in Congress Act to Block Shipments of Plutonium Here," *New York Times*, April 3, 1975; Salzman, Lorna, "New York Report: Plutonium," *Not Man Apart*, April 1975.

Page 168:
(Nader testimony on dangers of nuclear transportation) Rosenthal, Bruce, "Radioactive Wastes Widely Transported," *Critical Mass*, June 1975.
(far-flung nuclear shipments) Fialka, John, "Private Plutonium Business Is Bringing Big Money to the South," *Washington Star-News*, April 23, 1975.
(transporting waste from Peach Bottom) "M & P Tracks Eyed to Ship Nuclear Fuel," *York Daily Record*, April 1, 1975; letter from Ray Hovis, July 7, 1975.

Page 169:
(Barnwell shipments) Spake, "South Carolina's Silent Death Factory."
(Stroudsburg, Pa., accident) Anderson, Marion, "Fallout Along the Freeway: The Hazards of Transporting Radioactive Wastes in Michigan," Public Interest Research Group in Michigan report, Jan. 1974.

Page 170:
(Gleason accident) Sneddon, James D., "The Incredible Story of the Leaking Plutonium," *Beaver County* (Pa.) *Times*, Jan. 2, 1975.
(Rosenbaum report) "The Threat of Nuclear Theft and Sabotage," the Rosenbaum safeguards reports, *Congressional Record*, April 30, 1974, excerpts submitted by Senator Ribicoff.

Page 171:
(MITRE report) Warden, Rob, "Fear A-Plant Terror Raids" and "U.S. Agency Suppresses Report on Atomic Industry Protection," *Chicago Daily News*, Nov. 28, 1975.
(GAO criticism) "Improvements Needed in the Program for the Protection of Special Nuclear Material," GAO Report B-164105, Nov. 7, 1973; "Protecting Special Nuclear Material in Transit: Improvements Made and Existing Problems," GAO Report B-164105, April 12, 1964.

Page 172:
(Ribicoff, Ikle statements) statement, Senators Abe Ribicoff and Charles H. Percy, Senate Committee on Government Operations, April 29, 1975.
(Peterson statement) Baer, John, "Peterson Wants Nuclear Plants Built Off-Shore," *Harrisburg Evening News*, Feb. 21, 1974.
(terrorist groups) "Terrorism: Death in Rome Aboard Flight 110," *Time*, Dec. 31, 1973; Conine, Ernest, "Nuclear Terrorism: A Danger to Millions in U.S.," *Los Angeles Times*, Nov. 1, 1974; "Possibility

of Attempted Nuclear Thefts Causing Deep Concern," *Los Angeles Times*, Nov. 11, 1974, from Thomas O'Toole story in the *Washington Post*.

Page 173:
(Orlando, Fla., threat) Ingram, Timothy H., "Nuclear Hijacking: Now Within the Grasp of Any Bright Lunatic," *Washington Monthly*, Jan. 1973; Lapp, Ralph, "The Ultimate Blackmail," *New York Times Magazine*, Feb. 4, 1973.
(Con Ed arson) DeNike, Douglas L., "Radioactive Malevolence," *Bulletin of the Atomic Scientists*," reprint, 1974; Fialka, John, "Everything Went Wrong," *Washington Star-News*, April 24, 1974.
(Oak Ridge threat) "Hijackers Threaten Oak Ridge, Nuclear Facility Shuts Down," *Columbia* (S.C.) *Record*, Nov. 11, 1972.
(AEC spokesman) former AEC chairman James Schlesinger, on "Meet the Press," Dec. 17, 1972, from *Congressional Record— Senate*, Oct. 12, 1973.
(Argentine guerrillas) "Nuclear Terrorism," Public Interest Report, Environmental Alert Group, Los Angeles, 1974.
(Commonwealth Edison threat) "Nuclear Power Plant Gets Bomb Threats," AP story, *Janesville* (Wis.) *Gazette*, Sept. 3, 1974.
(French A-plant blast) "Bombs Hit Nuclear Site in France," *Washington Post*, May 4, 1974.

Page 174:
(underwater demolition) testimony by Dr. Bruce L. Welch before Ribicoff hearings, March 12–13, 1974, reprinted in *Congressional Record—Senate*, March 21, 1974.
(cutting safeguards funds) Olson, M. C., "Hot River Valley"; Conine, Ernest, "Nuclear Terrorism: A Danger to Millions in U.S.," *Los Angeles Times*, Nov. 1, 1974; "AEC Gives Plan for Tight Security at Nuclear Plants," *New York Times*, Nov. 8, 1974.
(Taylor security estimates) Wisby, Gary, "Moderate Price Put on Safeguards in Nuclear Thefts," *Chicago Sun-Times*, Sept. 12, 1975.

Page 175:
(Willrich quote) Willrich, Mason, "Terrorists Keep Out!" *Bulletin of the Atomic Scientists*, May 1975.
(Met Ed ex-guards) news release, June 4, 1975; Serbell, John, "Three Miles Island: 'Sabotage Would Be Easy,'" *Harrisburg Independent Press*, June 13–20, 1975.

Page 176:
(DeNike's San Onofre "demonstration") "Sedro Woolley N-Plant Called 'Sitting Duck,'" *Seattle Post-Intelligencer*, June 25, 1975.
(nuclear parks) Shorbe, Bill, "NRC Conducting 'Nuclear Parks' Study," *Not Man Apart*, July 1975; "Conceptualized Description of Nuclear Energy Centers," NRC, July 10, 1975; Draeger, Harlan, "Atom Clusters Would Make Giant Zion Plant a 'Midget,'" *Chicago Daily News*, Aug. 9, 1975; Jones, Clayton, "Giant U.S. 'Energy Parks': Dream Plan, or Nightmare?" *Christian Science Monitor*, Oct. 9, 1975.

Page 177:
(Taylor estimate of how much nuclear material needed for bomb)
Taylor, Theodore B., and Willrich, Mason, "Nuclear Theft: Risks
and Safeguards," Ballinger Publishing Co., Cambridge, Mass., 1974.
(Taylor Senate testimony) statement, July 15, 1974, before the Subcommittee on International Finance of the Committee on Banking,
etc.

Page 178:
(*New Yorker* series) McPhee, John, "The Curve of Binding Energy,"
parts 1–3, *New Yorker,* Dec. 3, 10, 17, 1973.

Page 179:
("The Plutonium Connection") presented March 9, 1975; Senator
Ribicoff's statement, *Congressional Record—Senate,* March 11, 1975;
Hemstock, Jack, "Now It's Easy to Make Your Own A-Bomb,"
Chicago Tribune, March 7, 1975; "Student Designs Atom Bomb in
Five Weeks," *Not Man Apart,* March 1975.
(missing nuclear material) "Possibility of Attempted Nuclear Thefts
Causing Deep Concern," *Los Angeles Times,* Nov. 10, 1974; Burn-
ham, David, "Thousands of Pounds of Materials Used in Nuclear
Bombs Are Unaccounted for," *New York Times,* Dec. 29, 1974;
Duran, Paco, "Missing Nuclear Materials Plague NRC, Industry,"
Critical Mass, Sept. 1975.
(price of plutonium) Shapley, Deborah, "Plutonium: Reactor Pro-
liferation Threatens a Nuclear Black Market," *Science,* April 9, 1971;
Kinderman, E. M., "Plutonium: Home Made Bombs?" statement,
Atomic Industrial Forum, March 6, 1972; Ingram, "Nuclear Hi-
jacking: Now Within the Grasp of Any Bright Lunatic."

Page 180:
(Larson quote) Larson, Clarence E., "Nuclear Materials Safeguards:
A Joint Industry Government Mission," address delivered at AEC
symposium in Los Alamos, Oct. 27–29, 1969.
(Mafia connections in shipping) Deborah Shapely's article in *Science,*
April 9, 1971.
(pilferage) "Worker Pilferage Huge," UPI story in *Chicago Tribune,*
Aug. 23, 1975.
(theft of military material) Coates, James, "Army Links Organized
Criminals to Worldwide Theft of Weapons," *Chicago Tribune,* Aug.
21, 1975.
(Muntzing quote) from David Burnham's article on missing nuclear
materials, *New York Times,* Dec. 29, 1974.

Page 181:
(nuclear security risks) "Nuclear Terrorism," Environmental Alert
Group, Los Angeles, 1974; DeNike, "Radioactive Malevolence,"
Bulletin of the Atomic Scientists, 1974.
(nuclear safeguards and civil liberties) Speth, Gus, Tamplin, Arthur,
and Cochran, Thomas, "Plutonium Recycle or Civil Liberties? We
Can't Have Both," *Environmental Action,* Dec. 7, 1974.
(Pomeroy case) "Texas Agency Destroys Disputed Files," *New
York Times,* Aug. 25, 1974.

(Nader) Nader, Ralph, "Private Police for Utilities," column, *Harrisburg Independent Press*, Jan. 31–Feb. 7, 1975.

Page 182:
(Comey) interview with Comey, Oct. 1974.
(Russell Ayres "demonstration") "Nuclear Safeguards Threaten Civil Liberties," *Critical Mass*, Dec. 1975; Comey, David D., "The Perfect Trojan Horse: Nuclear Power May Prove to Be Our Ultimate Internal Subversion," address to the American Law Institute and American Bar Association Conference on Atomic Energy, Washington, D.C., Dec. 11, 1975.

Page 183:
(nuclear proliferation) Lerch, Irving A., "Even the Meek Can Have a Mushroom-Shaped Club," *Chicago Tribune*, Sept. 1, 1974; Kotulak, Ronald, "Doomsday Clock Moves Toward Midnite: Experts," *Chicago Tribune*, Aug. 26, 1974; "Little Nations with Nuclear Ambitions," *New York Times*, April 13, 1975; Pomerance, Jo, "Meeting the Nuclear Threats," *Nation*, Aug. 31, 1974; Krieger, David, "The Dangers of Deterrence," *Progressive*, Oct. 1974; Childs, Marquis, "Bottling the Atomic Genie," column, *Washington Post*, March 25, 1975; "Brazil Contract for German A-Technology Stirs Concern," *Chicago Sun-Times*, June 1, 1975; "Nuclear Power: The High-Priced Spread," editorial, *Washington Post*, June 8, 1975; Feld, Bernard T., "Making the World Safe for Plutonium," editorial, *Bulletin of the Atomic Scientists*, May 1975; "Vienna Panel Proposes Stricter Guarding of Nuclear Materials," *New York Times*, May 3, 1975; Krieger, David, "Terrorists and Nuclear Technology," *Bulletin of the Atomic Scientists*, June 1975; "How Israel Got the Bomb," *Time*, April 12, 1976.

Page 184:
(Hahn quote) Hahn, Walter F., "Nuclear Proliferation," *Strategic Review*, vol. 3, Winter 1975.
(Ribicoff and Percy quotes) Phillips, James G., "Energy Report: Controversy Surrounds Proposed Nuclear Export Policies," *National Journal Reports*, May 10, 1975.

Page 185:
(Alfvén quote) "Energy and Environment," *Bulletin of the Atomic Scientists*," May 1972.

CHAPTER VII: The Dollar Goes Up

Pages 186 ff.:
(nuclear-plant reliability) Ehrich, Thomas, "Atomic Lemons: Breakdowns and Errors in Operation Plague Nuclear Power Plants," *Wall Street Journal*, May 3, 1973; "Uneconomical A-Plants," *(Honolulu) Star-Bulletin & Advertiser*, June 23, 1973, a *Christian Science Monitor* story; Comey, David D., "Nuclear Power Plant Reliability: A

Report to the Federal Energy Administration," Sept. 10, 1974; Draeger, Harlan, "Charge Nuclear Plants Perform at Only 54%," *Chicago Daily News,* Sept. 10, 1974; Draeger, Harlan, "U.S. Pollution Aide Hits A-Plant Production Lag," *Chicago Daily News,* Sept. 11, 1974; Bukro, Casey, "Economics of Nuclear Power Hit," *Chicago Tribune,* Sept. 29, 1974; "Critic Says Atom-Power Reliability Gap Could Cost $100 Billion in Next 15 Years," *Wall Street Journal,* Oct. 8, 1974; Olson, M. C., "Nuclear Energy: It Costs Too Much," *Nation,* Oct. 12, 1974; Young, Robert, "Edison Seeks Full Operation of Zion Nuclear Facility," *Chicago Tribune,* Dec. 5, 1974; *BPI Newsletter,* Spring 1975; Donsimoni, Marie-Paule, Treitel, Robert, and Bupp, Irvin C., "The Emonomics of Nuclear Power," *Technology Review,* Feb. 1975; Comey, David D., "Nuclear Power Plant Reliability: The 1973–74 Record," *Not Man Apart,* April 1975; Margen, Peter, and Lindhe, Soren, "The Capacity of Nuclear Power Plants," *Bulletin of the Atomic Scientists,* Oct. 1975—and Comey's articles, the same issue: "On Cooking Curves," "Points vs. Trends," and "Following the Leader."

Page 187:
(why Comey got interested in nuclear economics) address, Critical Mass conference, Washington, D.C., Nov. 16, 1975.

Page 189:
(Roddis report on Con Ed repairs) *Environmental Action Bulletin,* Dec. 9, 1972.
(Con Ed's safety troubles) Fialka, John, "Everything Went Wrong," *Washington Star-News,* April 24, 1974.
(Con Ed's failure to pay dividend) Margolis, Jon, and Kelly, Harry, "Electric Utility Business: Even When It's Good It's Bad," *Chicago Tribune,* May 26, 1975.

Page 190:
(New Yorkers' electric bills) Margolis, Jon, and Kelly, Harry, "Why the Power Cost Controversy Is Heating Up," *Chicago Tribune,* May 25, 1975.
(Ayers says it cost 22 percent less to generate nuclear electricity) Poulos, Nick, "Electric Utilities' Outlook Brightens," *Chicago Tribune,* July 23, 1975.

Page 191:
(Comey's report on reliability of large nuclear plants) Comey, David D., "Future Performance of Large Nuclear Plants," BPI report 7533, June 19, 1975.

Page 192:
(NRC report, Rasmussen statement) Burnham, David, "Federal Study Charges Little Concern by Utilities in Reliability of Reactors," *New York Times,* March 9, 1975.
(Heronemus testimony) reprinted, *Congressional Record,* May 15, 1973.
(rising costs of nuclear plants) *BPI Newsletter,* Spring 1975.

Page 193:
(original cost of Fulton plant) Olson, M. C., "Hot River Valley," *Nation,* Aug. 3, 1974.
(Anthony Bournia: new cost of Fulton plant; rising kilowatt cost) phone interview, July 1975.
(Strauss quote) Milius, Peter, "Major Battle Is Brewing on A-Plants," *Washington Post,* June 23, 1975.
(Chase Manhattan spokesman) Genachte, Paul F., "Moving Ahead with the Atom," Jan. 1957.
(Investors Responsibility report) "Study Clouds Nuclear Capital," *Not Man Apart,* mid-March 1975.

Page 194:
(Westinghouse cost estimate) Simpson, John W., "Nuclear Fission: Its Uses and Economics," paid position paper, *Fortune,* Feb. 1975.
(Friends of the Earth cost estimate) "Study Clouds Nuclear Capital," *Not Man Apart,* mid-March 1975.
(Zarb quote) address to Commonwealth Club of San Francisco, July 11, 1975, reported in *Not Man Apart,* Aug. 1975.

Page 195:
(Robert Smith estimate) Margolis, Jon, and Kelly, Harry, "Inefficiency Boosts Electric Bills, Critics Say," *Chicago Tribune,* May 28, 1975.
(Comey reply) interview, Jan. 1976.
(A-plant cancellations) "Five Utilities Delay Six Units, Cancel Another," *Wall Street Journal,* Sept. 10, 1974; Nordlinger, Stephen E., "Nuclear Plants Lose Favor," *Baltimore Sun,* March 31, 1974; "Utilities Doubt Nuclear Plants," *Critical Mass,* April 1975; Margolis and Kelly, *Chicago Tribune* article, May 28, 1975; "Reactor Sales Poor in 1975," *Not Man Apart,* Nov. 1975; Mateja, James, "Money Is Mightier Than Atom," *Chicago Tribune,* Dec. 7, 1975.
(halt of electric growth) same sources plus Casey Bukro's *Chicago Tribune* story of Sept. 29, 1974; "Utilities Report Summer Power Surplus," *Baltimore Sun,* May 15, 1975; "Energy Consumption Remains Stable," *Not Man Apart,* Aug. 1975.

Page 196:
(Simpson statement) Simpson, John W., "An Overview of Energy: The Need for a New Energy Base," paid position paper, *Fortune,* Jan. 1975.

Page 197:
(Gould quote) M. Olson's *Nation* story of Oct. 12, 1974.
(asking for rate hikes) Rosenberg, William G., "Utilities Need Help —Now!" opinion piece, *Wall Street Journal,* Jan. 8, 1975.
(rising electric bills) "The Bills Are Electrifying," *Newsweek,* April 8, 1974; Margolis and Kelly article, *Chicago Tribune,* May 25, 1975.
(Philadelphia Electric problems) Philadelphia Electric press release, Sept. 4, 1974; Brown, Warren, "PUC Opens Its Meeting, and Public Opens Fire" and "PE's Rates Up 63.6 Pct. in One Year," *Philadelphia Inquirer,* Aug. 25, 1974; "PUC Backs Off on Rate Increase," *Lancaster* (Pa.) *Intelligencer,* Oct. 1, 1974.

Pages 198–199:
(Vermont story) Kifner, John, "Shutdown of a Nuclear Plant, 17th in 19 Months, Spurs U.S. Debate," *New York Times*, March 31, 1974; "Vermont Passes Curb on A-Power," *New York Times*, April 1, 1975; "Vt. Controls Nuclear Power," *Critical Mass*, April 1975; "Vermont Nuclear Bill Signed by Governor Salmon," *Not Man Apart*, May 1975.

Pages 199–200:
(Consumers Power) from *Not Man Apart*: "Palisades to Pay Pipers," Dec. 1974; "The Jacks That Build Midland," Jan. 1975; "Pay the Devil for Midland One and Two," May 1975; plus Margolis and Kelly, *Chicago Tribune* article, May 28, 1975.

Pages 200 ff.:
(Jacksonville story) interviews by phone with Joe Cury, Oct., Dec. 1975, Jan. 1976, plus: "Joe Cury's Community Christmas Caper," *Florida Barb*, Dec. 6, 1973; "The Bills Are Electrifying," *Newsweek*, April 8, 1974; Clendinen, Dudley, "Electric Dollar Crunch Comes to Eden . . . and Grocer Is Jolted to Action," *St. Petersburg Times*, April 15, 1974; "Electric, Gas, Phone Bills Jump—Some Double or More," *U.S. News & World Report*, May 13, 1974; Ward, James R., "2 JEA Critics Ask Mayor to Fire Winnard, Ewton, Board," *Florida Times-Union*, Sept. 6, 1974; Clark, Mike, "Take High Power Bills to Higher Power," and Cearnal, Deborah, "Cury Urges Cut in JEA's Budget," *Florida Times-Union*, Aug. 10, 1974; Toner, Mike, "Floating Power Plants—Will the Dream Crumble?" *Miami Herald*, Oct. 27, 1974; Ruediger, Steve, "Group Fights Utility Rates," *Florida Times-Union*, Nov. 5, 1974; Clendinen, Dudley, "How 'POWER' Sank Boondoggle," *St. Petersburg Times*, Dec. 2, 1974; "Offshore Power: Jacksonville's Troubled Savior," *Florida Trend*, Jan. 1975; Ward, James R., "Westinghouse Gets Sole OPS Control: Tenneco's Pullout Held No Threat," *Florida Times-Union*, Feb. 1, 1975; Brown, Lloyd, "Candidate Cites Campaign 'Dirty Tricks,'" *Jacksonville Journal*, April 2, 1975; "FBI Is Investigating Threat on Cury's Life," *Florida Times-Union*, April 3, 1975; Markowitz, Jack, "WE to U.S.: Buy 8 Floating Power Plants Now," *Pittsburgh Post-Gazette*, May 1, 1975; Clendinen, Dudley, "Jacksonville's Giant Business Gamble," *St. Petersburg Times*, June 8, 1975; Clendinen, Dudley, "Jacksonville Now May Get Overdue Facts," *St. Petersburg Times*, Aug. 25, 1975; "A City That Reached for Riches and Got Headaches Instead," *U.S. News & World Report*, Sept. 1, 1975; "Joe Cury's Claims Against Staten Published in St. Pete," *Jacksonville Journal*, Dec. 22, 1975; Clendinen, Dudley, "Ford Fund Raiser Figures in Probe," *St. Petersburg Times*, Dec. 22, 1975.

Pages 205–206:
(Rancho Seco) Harding's testimony: April 15, 1975; Walbridge, William C., "Report on Future Generation," Jan. 8, 1976; Nader's remark: Critical Mass conference, Washington, D.C., Nov. 16, 1975.

CHAPTER VIII: The Citizens' Battle

Pages 207–222:
(unless otherwise specified) Olson, M. C., "The Hot River Valley," *Nation,* Aug. 3, 1974.

Page 210:
(Nader quote) Dick Cavett TV show, Dec. 13, 1972, reprinted in the *Congressional Record,* March 15, 1973.
(*New York Times* editorial) "Research for Power," Jan. 31, 1973.

Pages 213 ff.:
(court ruling on Peach Bottom) U.S. Court of Appeals, District of Columbia, No. 74–1923, *York Committee for a Safe Environment, et al.,* v. *U.S. NRC and Philadelphia Electric Co.,* argued Nov. 3, 1975; decided, Dec. 9, 1975.

Page 214:
(Gofman) Gofman, John W., "Time for a Moratorium," *Environmental Action,* Nov. 1972; Gofman letter, Nov. 28, 1974.

Page 221:
(press reports of nuclear mishaps) "22 Violations Cited at 3-Mile Island," *Lancaster* (Pa.) *New Era,* Sept. 13, 1974; "AEC Closes Two York Atomic Units," *Lancaster* (Pa.) *Intelligencer Journal,* Sept. 23, 1974; "N-Reactor Shut Down at Peach Bottom Unit," *Lancaster* (Pa.) *Intelligencer Journal,* Oct. 18, 1974; "N-Power Station Unit Will Reopen," *Harrisburg Patriot,* Oct. 25, 1974; Kessler, Charles H., "Radioactive Gas Leaks at A-Plant," *Lancaster* (Pa.) *New Era,* Oct. 15, 1974; Baer, John, "3 'Minor' Leaks of Radioactive Gas 'Not Told' to State," *Harrisburg Evening News,* Oct. 17, 1974; "N-Plant Shut Down Due to Leak of Gas," *Lancaster* (Pa.) *Intelligencer Journal,* Oct. 24, 1974; "Power Unit Shut Down at N-Plant," *Harrisburg Patriot,* Nov. 20, 1974; "'Radioactive Gases Released at N-Facility, Met Ed Admits," *Harrisburg Patriot,* Jan. 28, 1975.

Page 222:
(Fulton A-plant backdown) Shaw, Charles, "Economy Stalls Two N-Plants," *Lancaster* (Pa.) *Intelligencer Journal,* Sept. 5, 1975; "Utility Urged to Scrap County N-Plant Plans," *Lancaster* (Pa.) *Intelligencer Journal,* Oct. 11, 1974; "Farmers Oppose Nuclear Plants," *Lancaster* (Pa.) *Intelligencer Journal,* Oct. 18, 1974; Donalson, Al, "Need, Safety of A-Energy Questioned," *Pittsburgh Press,* April 13, 1975; "GA Abandons Commercial Reactors," *Not Man Apart,* Dec. 1975.

Pages 223 ff.:
(Koshkonong A-plant battle: extensive interviews in Wisconsin;

coverage of the PSC hearings on the Koshkonong plant; plus the following) Lovejoy, Steven T., "Koshkonong Nuclear Plant Spurs Squabble," *Wisconsin State Journal,* Aug. 2, 1974; Ingersoll, Bruce, "Foes of A-Plant Hit Financing," *Chicago Sun-Times,* Aug. 3, 1974; Bukro, Casey, "AEC Has Critic as Adviser," *Chicago Tribune,* Aug. 13, 1974; Rowen, James, "Madison Says 'No,'" *Nation,* Aug. 17, 1974; Legro, Ron, "How Much Extra Power Needed," *Milwaukee Sentinel,* Aug. 21, 1974; Jordan, William, III, "PSC Restricts More Spending for Atom Plant," *Wisconsin State Journal,* Aug. 22, 1974; Legro, Ron, "Pacts on A-Plant Halted," *Milwaukee Sentinel,* Aug. 22, 1974; Legro, Ron, "A-Plant Safety Replies Blocked," *Milwaukee Sentinel,* Aug. 23, 1973; "Nuclear Safety Jurisdictional Fight," *Jefferson Daily County Union,* Aug. 26, 1974; "Lucey's' Criticism Attacked by Utility," and Kienitz, Richard C., "PSC Checks Electric Co. Projections," *Milwaukee Journal,* Aug. 28, 1974; Jordan, William R., III, "Nader May Join Fight on A-Plant," *Wisconsin State Journal,* Aug. 30, 1974; Legro, Ron, "Cherry Stirs Fuss on A-Plant Issue," *Milwaukee Sentinel,* Aug. 26, 1974; "Aspin Urges PSC to Make Nuclear Injunction Permanent," and "Nuclear Safety Raised as Hearing Continues," *Janesville* (Wis.) *Gazette,* Aug. 27, 1974; Gould, Whitney, "Cherry Takes Aim at A-Plant," *Madison Capital Times,* Aug. 28, 1974; Banaszynski, Jacqui, "Opponent Quizzes Power Firms on Expansion of Nuclear Plant," *Janesville Gazette,* Aug. 28, 1974; "A-Plants and the Public," editorial, *Madison Capital Times,* Aug. 29, 1974; "WP & L Seeking Permission to Boost Its Rates Again," *Milwaukee Sentinel,* Aug. 30, 1974; Legro, Ron, "How Much Extra Power Needed, Utilities Asked," *Milwaukee Sentinel,* Aug. 29, 1974; Banaszynski, Jacqui, "Nuclear Plant Debate Heats Up," *Janesville Gazette,* Aug. 30, 1974; Rowen, James, "Lake Koshkonong A-Plant: The People Will Pay," *Daily Cardinal,* fall registration issue, 1974; Kienitz, Richard C., "Complex Questions on A-Plant," *Milwaukee Journal,* Sept. 1, 1974; "Over a Billion Already Pledged," *Jefferson Banner,* Sept. 4, 1974; O'Neill, Tom, "Public Will Get Short Shrift on Nuclear Plant, Lucey Says," *Janesville Gazette,* Sept. 4, 1974; "Nuclear Issue Packs Township Meeting; Most Oppose Project," *Jefferson County Daily Union,* Sept. 6, 1974; Lee, Thomas, "New A-Plant Is Given Heat by Residents," *Wisconsin State Journal,* Sept. 6, 1974; "Cherry Says PSC Should Get Tougher with Utilities," *Janesville Gazette,* Sept. 6, 1974; "Utility Expenditure Could Tie PSC Hands," editorial, *Milwaukee Journal,* Sept. 9, 1974; "River and Lake Association Asks Nuclear Plant Questions," *Jefferson County Daily Union,* Sept. 10, 1974; Legro, Ron, "Battle on A-Plant Has Only Begun," *Milwaukee Sentinel,* Sept. 3, 1974; Jaeger, Richard W., "Power Line Session Boycotted," *Wisconsin State Journal,* Sept. 13, 1974; Haug, John, "Pickets Protest East Dane Power Lines," *Madison Capital Times,* Sept. 13, 1974; Bjorklund, Robert C., "Farmers Irked at Survey Crew," *Wisconsin State Journal,* Sept. 21, 1974; "Nuclear Plant Site Opposed," *Wisconsin State Journal,* Sept. 25, 1974; "Nuclear Co-operation," editorial, *Wisconsin State Journal,* Sept. 26, 1974; Gould, Whitney, "PSC Newcomer is 'Nuclear Skeptic,'" *Madison Capital Times,* April 16, 1975; letter, "Koshkonong Nuclear Power Plant

Project," from Myron Cherry to Friends of the Earth, June 4, 1975; "Mike Cherry Wins Two Big Ones," *Not Man Apart,* July 1975; "Kosh Quashed," *Not Man Apart,* mid-Jan. 1975.

Pages 230 ff.:
(Critical Mass '74) my coverage of the conference and article, "New Protest Movement: Reacting to the Reactors," *Nation,* Jan. 11, 1975.

Page 233:
(pressures on Congress) "That Was the Year That Was: A Sorry One," Llewellyn King, editor, *Weekly Energy Report,* Dec. 23, 1974; Phillips, James C., "Nader, Nuclear Industry Prepare to Battle over the Atom," *National Journal Reports,* Feb. 1, 1975; Gillette, Robert, "Atomic Energy Will Be a Prime Source of Debate," *New York Times,* Feb. 9, 1975; Burnham, David, "Congress Faces Key Decisions on Nuclear Reactors," *New York Times,* March 30, 1975.
(pronuclear statement) "Energy Policy Statement," Jan. 16, 1975.

Page 234:
(antinuclear statement) "Declaration on the 30th Anniversary of Hiroshima," delivered to the White House, Aug. 6, 1975.

Pages 235–236:
(nuclear establishment counterattack) Phillips, James G., "Atomic Industry Drafts Plan for Major Lobbying Effort," *National Journal Reports,* April 4, 1975; "AIF: Madison Ave. PR," *Critical Mass,* April 1975; Milius, Peter, "Major Battle Is Brewing on A-Plants," *Washington Post,* June 23, 1975; Abbotts, John, "Industry Campaign Minimizes Plutonium Hazard," *Critical Mass,* July 1975; Galazen, Thomas, "The Nuclear Power Lobby," *Progressive,* July 1975; Hill, Gladwin, "Nuclear Power Development Encounters Rising Resistance with Curbs Sought in a Number of States," *New York Times,* July 29, 1975; "Industry Girds for Battle," *Not Man Apart,* Aug. 1975.

Page 236:
(Harris poll) Ziomek, Jon, "Poll Finds 63% Favoring More Nuclear Energy Generators," *Chicago Sun-Times,* Aug. 22, 1975.
(Critical Mass '75) *Weekly Energy Report,* Nov. 18, 1975; Spake, Amanda, "Critical Mass '75," *Progressive,* Jan. 1975; "Critical Mass '75 Conference Marks Increase in Citizen Activity," *Critical Mass,* Dec. 1975.

Page 237:
(Ford administration programs for nuclear industry) "Administration Moves to Help Utilities Build More Nuclear Plants," *Critical Mass,* April 1975; "Ford Weighs Energy Corporation," *Chicago Sun-Times,* Aug. 23, 1975; Laitner, Skip, "Bailing Out the Nuclear Industry: ERDA Pushes Atomic Fuel Plan to Minimize Industry Risks," *Critical Mass,* Feb. 1976.

(Price-Anderson Act extended) Knight, Jeffrey, "Congress in Action," *Not Man Apart,* Feb. 1976.
(Pollard) "Atomic Safety Regulator Quits," *Chicago Sun-Times,* Feb. 10, 1976; Constantine, Peggy, "Urges Shutting Nuclear Plants," *Chicago Sun-Times,* Feb. 22, 1976; Van, Jon, "Many Nuclear Units Unsafe: Engineer," *Chicago Tribune,* Feb. 22, 1976; Hines, William, "Full Probe of Nuclear Power Plant Safety Hinted," *Chicago Sun-Times,* Feb. 24, 1976.

Page 238:
(critic, "sinking ship") Martell, Edward A., phone interview, Feb. 1976.
(phaseout) Bukro, Casey, "Phase Out Atom Plants, Ex-Edison Aide Warns," *Chicago Tribune,* Feb. 13, 1976.

CHAPTER IX: The Future

Pages 239 ff.:
(geothermal power) "Geothermal Energy Looms as Economic Factor," *Oil Digest,* Jan. 1971; Handerson, J. Hilbert, "Geothermal Heat, Our Next Major Source of Power," address, American Institute of Chemical Engineers, Dallas, Jan. 1972; "California's Electricity Quandary: III. Slowing the Growth Rate," Rand Corporation report, Sept. 1972; "The Geysers," Pacific Gas and Electric Co.–Union Oil Co., Sept. 1974; "Union Oil Company Operations at the Geysers," Union Oil Company, 1974; Pickens, Tom, "What We Can Do to Keep the World from Going Dark," *Chicago Tribune,* April 13, 1975; "Considering the Alternatives," *Time,* April 14, 1975.

Page 241:
(ERDA budget) Abbotts, John, Laitner, Skip, and Wasilczyk, John, "ERDA R&D Plan Misrepresents Energy Capabilities," *Critical Mass,* Aug. 1975; "Energy Priorities of ERDA's Operating Budget," *Not Man Apart,* May 1975.
(Club of Rome) Mesarovic, Mihajlo, and Pestel, Eduard, "Mankind at the Turning Point: The Second Report to the Club of Rome," E. P. Dutton and Co./Reader's Digest Press, New York, 1974.

Page 243:
(Ford Foundation report) Cowan, Edward, "Study Asks Halving of Energy Growth; Foresees No Harm," *New York Times,* Oct. 18, 1974.
(AIA reports) "Energy and the Built Environment" and "A Nation of Energy Efficient Buildings by 1990," American Institute of Architects, 1975.

Page 244:
(Ross and Williams report) "U.S. Could Have Used 45 Percent Less Energy and Lived as High," *Not Man Apart,* mid-Jan. 1976.
(Dow Chemical report) "Waste Industrial Steam Could Generate Electricity to Spare," *Not Man Apart,* Nov. 1975.

(Lee Schipper) address, Critical Mass Conference, Washington, D.C., Nov. 16, 1975.

Page 245:
(coal) "Coal Research Unit to Be Reorganized," *New York Times,* Feb. 24, 1973; "Breakthrough Allows Use of 'Dirty' Coal," *Milwaukee Journal,* Sept. 25, 1974; "Sulfur Can be 'Scrubbed' from Coal Power Wastes," *Wisconsin State Journal,* Sept. 26, 1974.

Pages 245 ff.:
(solar power) "U.S. Needs Sun Power, Zarb Says," *Baltimore Sun,* May 15, 1975.
(solar-power background) Henry, Omer, "Solar Homes," *Washington Post*'s *Potomac* magazine, Sept. 26, 1971; Meinel, Aden Baker, and Meinel, Marjorie Pettit, "Is It Time for a New Look at Solar Energy?" and Ford, Norman C., and Kane, Joseph W., "Solar Power," *Bulletin of the Atomic Scientists,* Oct. 1971; Arnold, William F., "Will Solar Cells Shine on Earth?" *Electronics,* May 22, 1972; O'Connor, Egan, "A Sunshine Future or a Radioactive One," Committee for Nuclear Responsibility, April 1973; Senator Mike Gravel newsletter on solar power, May 22, 1973; Keyerleber, Karl, "Clean Power from the Sun," *Nation,* Oct. 29, 1973; Henahan, John F., "How Soon the Sun: A Timetable for Solar Energy," *Saturday Review/World,* Nov. 20, 1973; Cowan, Edward, "Scientists Opposed to Nuclear Power Accuse AEC of Misleading the Public About Solar Energy," *New York Times,* April 23, 1974; "Complex to Rely on Solar Energy," *Milwaukee Sentinel,* Aug. 28, 1974; Mike Gravel newsletter, Sept. 1974; "High Rise to Use Solar Heat," *Milwaukee Journal,* Sept. 1, 1974; "Technology Is Here: Solar Heat Looks Better as Fuel Costs Skyrocket," *Washington Star-News,* Sept. 13, 1974; "PP&L Unveils 'Energy House,'" *Lancaster* (Pa.) *Intelligencer Journal,* Oct. 17, 1974; Henderson, Randi, "Putting the Sun to Work," *Baltimore Sun,* Dec. 12, 1974; Barnes, Peter, "The Solar Derby," *New Republic,* Feb. 1, 1975; "New Energy Group Set Up," *Baltimore Sun,* April 22, 1975; Cowan, Edward, "Solar Energy Is Gaining as Future Fuel Solution," *New York Times,* June 1, 1975; Jones, John A., "Solar Power: More Than a Ray of Hope," *Los Angeles Times,* June 29, 1975; Laitner, Skip, "Legislators Promote Solar Energy Bills," *Critical Mass,* Aug. 1975; Hines, William, "U.S. Bares Solar Energy Program to Year 2020," *Chicago Sun-Times,* Aug. 14, 1975; "Energy Goal: 25 Percent Solar by 2020," *Science News,* Aug. 23, 1975; Gelb, Debbie, "Solar Heating and Cooling of Buildings," *Critical Mass,* Sept. 1975; von Hippel, Frank, and Williams, Robert H., "Solar Technologies," *Bulletin of the Atomic Scientists,* Nov. 1975; Alexander, George, "Here Comes the Sun," *Los Angeles Times* story by George Alexander in *Chicago Sun-Times,* Dec. 8, 1975; Furgurson, Ernest, "Town Going Solar," *Hammond* (Ind.) *Times,* Dec. 28, 1975; Banks, Nancy, "Revolutionary New Device Uses Sun for Heat," *Reader,* Jan. 9, 1976; Laitner, Skip, "Decentralized Energy Systems," *Critical Mass,* Jan. 1976; Laitner, "Photovoltaic Systems," *Critical Mass,* Feb. 1976; Shurcliff, William A., "Active-Type Solar Heating Systems for

Houses: A Technology in Ferment," *Bulletin of the Atomic Scientists,* Feb. 1976; Douglas, Anne, "Solar House: A Functional Eye Pleaser," *Chicago Tribune,* Feb. 28, 1976.

Pages 248–249:
(using garbage and refuse for energy) Sumner, Jim, "Cow Manure Eyed As Energy Source," *Lancaster* (Pa.) *Intelligencer Journal,* Aug. 31, 1973; "The Methane Digester," Pennsylvania State University Cooperative Extension Service, Jan. 28, 1974; Grout, A. Roger, "Methane Gas Generation from Manure," Commonwealth of Pennsylvania, Dept. of Agriculture, 1974; Appell, H. R., et al., "Converting Organic Wastes to Oil—a Replenishable Energy Source," U.S. Bureau of Mines report, 1971; Singer, Dale, "St. Louis Finding Power Plant Fuel in the City Dump," *Houston Chronicle,* Feb. 4, 1973; "Heat Your House with Garbage," *Chicago Today,* Nov. 19, 1972; "Pittsburgh's Garbage-to-Power Plant a Step Near Reality," *Save Solanco Environment Newsletter,* May 1975; Nash, Hugh, "Make Mine Methanol," *Not Man Apart,* April 1975.

Pages 249–250:
(fusion) "The Promise of Fusion," *PG&E Progress,* Feb. 1974; Dudziak, Donal J., and Krakowski, "A Comparative Analysis of D-T Fusion Reactor Radioactivity and Afterheat," Study for the Los Alamos Scientific Laboratory, under AEC contract, 1974; Weismantel, Guy E., "Fusion's Engineering Phase," *Chemical Engineering,* Aug. 19, 1974; Conn, Robert W., statement on "Controlled Fusion as an Advanced Energy System in U.S.," address, Project Independence Hearings, Chicago, Sept. 10, 1974; Dudley, H. C., "The Ultimate Catastrophe," *Bulletin of the Atomic Scientists,* Nov. 1975; "Big Advance Told in Taming Fusion Energy," *Chicago Sun-Times,* Nov. 6, 1975.

Page 250:
(wind power) Heronemus, William E., "The U.S. Energy Crisis: Some Proposed Gentle Solutions," paper presented Jan. 12, 1972, joint meeting of local sections of the American Society of Mechanical Engineers and the Institute of Electrical and Electronics Engineers, West Springfield, Mass.; Heronemus, "Pollution-Free Energy from Off-Shore Winds," published by Marine Technology Society, Sept. 1972; Heronemus, written testimony submitted, Aug. 17, 1973, hearings on nuclear power, Commonwealth of Pennsylvania, Dept. of Insurance; Inglis, David Rittenhouse, "Wind Power Now!" *Bulletin of the Atomic Scientists,* Oct. 1975.
(ocean thermal power) Douglass, Robert H., statement, "Ocean Thermal Energy Conversion," delivered June 5, 1975, to U.S. House Subcommittee on Energy and Environment; Shobe, Bill, "Tapping the Heat of Davy Jones' Locker," *Not Man Apart,* Aug. 1975; my phone interview with Robert Douglass, June 1975; Zener, Clarence, "Solar Sea Power," *Bulletin of the Atomic Scientists,* Jan. 1976.

Page 251:
(McCormack) address, University of Notre Dame, April 14, 1975,

and address, National Security Industrial Association, Washington, D.C., April 9, 1975.
(Gofman) interview, June 1975 in San Francisco, and his address, "Alice in Blunderland," Oct. 17, 1975.

Page 252:
(Sol Burstein) interview, Dec. 1975.
(Kendall) *Critical Mass,* Dec. 1975.

Acronyms and Initials

AEC	Atomic Energy Commission
AIA	American Institute of Architects
BPI	Business and Professional People for the Public Interest
EBR	experimental breeder reactor
ECCS	emergency core cooling system
EPA	Environmental Protection Agency
ERDA	Energy Research and Development Administration
FEA	Federal Energy Administration
GE	General Electric
IAEA	International Atomic Energy Agency
JEA	Jacksonville Electric Authority
MIT	Massachusetts Institute of Technology
MUF	materials unaccounted for
NRC	Nuclear Regulatory Commission
OPS	Offshore Power Systems
POWER	People Outraged with Electric Rates
PRDC	Power Reactor Development Company
PSC	Public Service Committee
TVA	Tennessee Valley Authority

APPENDIX
Nuclear Reactors in the United States

SITE	PLANT NAME	CAPACITY (Net Kilowatts)	UTILITY	COMMERCIAL OPERATION
ALABAMA				
Decatur	Browns Ferry Nuclear Power Plant: Unit 1	1,067,000	Tennessee Valley Authority	1974
Decatur	Browns Ferry Nuclear Power Plant: Unit 2	1,067,000	Tennessee Valley Authority	1975
Decatur	Browns Ferry Nuclear Power Plant: Unit 3	1,067,000	Tennessee Valley Authority	1976
Dothan	Joseph M. Farley Nuclear Plant: Unit 1	829,000	Alabama Power Co.	1976
Dothan	Joseph M. Farley Nuclear Plant: Unit 2	829,000	Alabama Power Co.	1977
Clanton	Alan R. Barton Nuclear Plant: Unit 1	1,159,000	Alabama Power Co.	1985
Clanton	Alan R. Barton Nuclear Plant: Unit 2	1,159,000	Alabama Power Co.	1986
Scottsboro	Bellefonte Nuclear Plant: Unit 1	1,213,000	Tennessee Valley Authority	1980
Scottsboro	Bellefonte Nuclear Plant: Unit 2	1,213,000	Tennessee Valley Authority	1981
ARIZONA				
Wintersburg	Palo Verde Nuclear Generating Station: Unit 1	1,237,700	Arizona Public Service	1982
Wintersburg	Palo Verde Nuclear Generating Station: Unit 2	1,237,700	Arizona Public Service	1984
Wintersburg	Palo Verde Nuclear Generating Station: Unit 3	1,237,700	Arizona Public Service	1986
ARKANSAS				
Russellville	Arkansas Nuclear One: Unit 1	850,000	Arkansas Power & Light Co.	1974
Russellville	Arkansas Nuclear One: Unit 2	312,000	Arkansas Power & Light Co.	1977
CALIFORNIA				
Eureka	Humboldt Bay Power Plant: Unit 3	65,000	Pacific Gas and Electric Co.	1963
San Clemente	San Onofre Nuclear Generating Station: Unit 1	430,000	So. Calif. Ed. & San Diego Gas & El. Co.	1968
San Clemente	San Onofre Nuclear Generating Station: Unit 2	1,100,000	So. Calif. Ed. & San Diego Gas & El. Co.	1981
San Clemente	San Onofre Nuclear Generating Station: Unit 3	1,100,000	So. Calif. Ed. & San Diego Gas & El. Co.	1982
Diablo Canyon	Diablo Canyon Nuclear Power Plant: Unit 1	1,084,000	Pacific Gas and Electric Co.	1976
Diablo Canyon	Diablo Canyon Nuclear Power Plant: Unit 2	1,106,000	Pacific Gas and Electric Co.	1977
Clay Station	Rancho Seco Nuclear Generating Station	913,000	Sacramento Municipal Utility District	1975
*	—	1,128,000	Pacific Gas & Electric Co.	1981
*	—	1,128,000	Pacific Gas & Electric Co.	1982
Blythe	Sundesert Nuclear Plant: Unit 1	950,000	San Diego Gas & Electric Co.	1985
Blythe	Sundesert Nuclear Plant: Unit 2	950,000	San Diego Gas & Electric Co.	1987
COLORADO				
Platteville	Ft. St. Vrain Nuclear Generating Station	330,000	Public Service Co. of Colorado	1975

SITE	PLANT NAME	CAPACITY (Net Kilowatts)	UTILITY	COMMERCIAL OPERATION
CONNECTICUT				
Haddam Neck	Haddam Neck Plant	575,000	Conn. Yankee Atomic Power Co.	1968
Waterford	Millstone Nuclear Power Station: Unit 1	652,100	Northeast Nuclear Energy Co.	1971
Waterford	Millstone Nuclear Power Station: Unit 2	828,000	Northeast Nuclear Energy Co.	1975
Waterford	Millstone Nuclear Power Station: Unit 3	1,156,000	Northeast Nuclear Energy Co.	1979
DELAWARE				
Summit	Summit Power Station: Unit 1	770,000	Delmarva Power & Light Co.	1981
Summit	Summit Power Station: Unit 2	770,000	Delmarva Power & Light Co.	1984
FLORIDA				
Florida City	Turkey Point Station: Unit 3	666,000	Florida Power & Light Co.	1972
Florida City	Turkey Point Station: Unit 4	666,000	Florida Power & Light Co.	1973
Red Level	Crystal River Plant: Unit 3	825,000	Florida Power Corp.	1976
Ft. Pierce	St. Lucie Plant: Unit 1	810,000	Florida Power & Light Co.	1975
Ft. Pierce	St. Lucie Plant: Unit 2	810,000	Florida Power & Light Co.	1980
South Dade	—	1,100,000	Florida Power & Light Co.	1983
South Dade	—	1,100,000	Florida Power & Light Co.	1985
GEORGIA				
Baxley	Edwin I. Hatch Nuclear Plant: Unit 1	786,000	Georgia Power Co.	1975
Baxley	Edwin I. Hatch Nuclear Plant: Unit 2	795,000	Georgia Power Co.	1978
Waynesboro	Alvin W. Vogtle, Jr. Plant: Unit 1	1,113,000	Georgia Power Co.	—
Waynesboro	Alvin W. Vogtle, Jr. Plant: Unit 2	1,113,000	Georgia Power Co.	—
ILLINOIS				
Morris	Dresden Nuclear Power Station: Unit 1	200,000	Commonwealth Edison Co.	1960
Morris	Dresden Nuclear Power Station: Unit 2	809,000	Commonwealth Edison Co.	1970
Morris	Dresden Nuclear Power Station: Unit 3	809,000	Commonwealth Edison Co.	1971
Zion	Zion Nuclear Plant: Unit 1	1,050,000	Commonwealth Edison Co.	1973
Zion	Zion Nuclear Plant: Unit 2	1,050,000	Commonwealth Edison Co.	1974
Cordova	Quad-Cities Station: Unit 1	800,000	Comm. Ed. Co.—Ia.—Ill. Gas & Elec. Co.	1972
Cordova	Quad-Cities Station: Unit 2	800,000	Comm. Ed. Co.—Ia.—Ill. Gas & Elec. Co.	1972
Seneca	LaSalle County Nuclear Station: Unit 1	1,078,000	Comm. Ed. Co.—Ia.	1978
Seneca	LaSalle County Nuclear Station: Unit 2	1,078,000	Comm. Ed. Co.—Ia.	1979
Byron	Byron Station: Unit 1	1,120,000	Comm. Edison Co.	1980

SITE	PLANT NAME	CAPACITY (Net Kilowatts)	UTILITY	COMMERCIAL OPERATION
Byron	Byron Station: Unit 2	1,120,000	Comm. Edison Co.	1982
Braidwood	Braidwood: Unit 1	1,120,000	Comm. Edison Co.	1981
Braidwood	Braidwood: Unit 2	1,120,000	Comm. Edison Co.	1982
Clinton	Clinton Nuclear Power Plant: Unit 1	933,400	Illinois Power Co.	1981
Clinton	Clinton Nuclear Power Plant: Unit 2	933,400	Illinois Power Co.	1984
INDIANA				
Westchester	Bailly Generating Station	645,300	Northern Indiana Public Service Co.	—
Madison	Marble Hill Nuclear Power Station: Unit 1	1,130,000	Public Service Indiana	1982
Madison	Marble Hill Nuclear Power Station: Unit 2	1,130,000	Public Service Indiana	1984
IOWA				
Palo	Duane Arnold Energy Center: Unit 1	535,000	Iowa Electric Light and Power Co.	1975
KANSAS				
Burlington	Wolf Creek Generation Station: Unit 1	1,150,000	Kansas Gas & Electric—Kansas City P & L	1982
LOUISIANA				
Taft	Waterford Generating Station: Unit 3	1,113,000	Louisiana Power & Light Co.	1980
St. Francisville	River Bend Station: Unit 1	934,000	Gulf States Utilities Co.	1981
St. Francisville	River Bend Station: Unit 2	934,000	Gulf States Utilities Co.	1983
MAINE				
Wiscasset	Maine Yankee Atomic Power Plant	790,000	Maine Yankee Atomic Power Co.	1972
Sears Island	—	1,150,000	Maine Yankee Atomic Power Co.	—
MARYLAND				
Lusby	Calvert Cliffs Nuclear Power Plant: Unit 1	845,000	Baltimore Gas and Electric Co.	1975
Lusby	Calvert Cliffs Nuclear Power Plant: Unit 2	845,000	Baltimore Gas and Electric Co.	1977
Douglas Point	Douglas Point Project Nuclear Gen. Station: Unit 1	1,178,000	Potomac Electric Power Co.	1985
Douglas Point	Douglas Point Project Nuclear Gen. Station: Unit 2	1,178,000	Potomac Electric Power Co.	1987
MASSACHUSETTS				
Rowe	Yankee Nuclear Power Station	175,000	Yankee Atomic Electric Co.	1961
Plymouth	Pilgrim Station: Unit 1	670,000	Boston Edison Co.	1972
Plymouth	Pilgrim Station: Unit 2	1,180,000	Boston Edison Co.	1982
Montague	—	1,150,000	Northeast Utilities	—
Montague	—	1,150,000	Northeast Utilities	—

SITE	PLANT NAME	CAPACITY (Net Kilowatts)	UTILITY	COMMERCIAL OPERATION
MICHIGAN				
Big Rock Point	Big Rock Point Nuclear Plant	75,000	Consumers Power Co.	1965
South Haven	Palisades Nuclear Power Station	700,000	Consumers Power Co.	1971
Lagoona Beach	Enrico Fermi Atomic Power Plant: Unit 2	1,093,000	Detroit Edison Co.	1980
Bridgman	Donald C. Cook Plant: Unit 1	1,060,000	Indiana & Michigan Electric Co.	1975
Bridgman	Donald C. Cook Plant: Unit 2	1,060,000	Indiana & Michigan Electric Co.	—
Midland	Midland Nuclear Power Plant: Unit 1	458,000	Consumers Power Co.	1982
Midland	Midland Nuclear Power Plant: Unit 2	808,000	Consumers Power Co.	1981
St. Clair County	Greenwood: Unit 2	1,200,000	Detroit Edison Co.	1984
St. Clair County	Greenwood: Unit 3	1,200,000	Detroit Edison Co.	1986
MINNESOTA				
Monticello	Monticello Nuclear Generating Plant	545,000	Northern States Power Co.	1971
Red Wing	Prairie Island Nuclear Generating Plant: Unit 1	530,000	Northern States Power Co.	1973
Red Wing	Prairie Island Nuclear Generating Plant: Unit 2	530,000	Northern States Power Co.	1975
MISSISSIPPI				
Port Gibson	Grand Gulf Nuclear Station: Unit 1	1,250,000	Mississippi Power & Light Co.	1979
Port Gibson	Grand Gulf Nuclear Station: Unit 2	1,250,000	Mississippi Power & Light Co.	1984
MISSOURI				
Fulton	Callaway Plant: Unit 1	1,120,000	Union Electric Co.	1981
Fulton	Callaway Plant: Unit 2	1,120,000	Union Electric Co.	1983
NEBRASKA				
Fort Calhoun	Ft. Calhoun Station: Unit 1	457,400	Omaha Public Power District	1973
Fort Calhoun	Ft. Calhoun Station: Unit 2	1,136,000	Omaha Public Power District	1983
Brownville	Cooper Nuclear Station	778,000	Nebraska Public Power District and Iowa Power and Light Co.	1974
NEW HAMPSHIRE				
Seabrook	Seabrook Nuclear Station: Unit 1	1,200,000	Public Service of N.H.	1980
Seabrook	Seabrook Nuclear Station: Unit 2	1,200,000	Public Service of N.H.	1982

SITE	PLANT NAME	CAPACITY (Net Kilowatts)	UTILITY	COMMERCIAL OPERATION
NEW JERSEY				
Forked River	Oyster Creek Nuclear Plant: Unit 1	640,000	Jersey Central Power & Light Co.	1969
Forked River	Forked River Generating Station: Unit 1	1,070,000	Jersey Central Power & Light Co.	1982
Salem	Salem Nuclear Generating Station: Unit 1	1,090,000	Public Service Electric and Gas, N.J.	1976
Salem	Salem Nuclear Generating Station: Unit 2	1,115,000	Public Service Electric and Gas, N.J.	1979
Salem	Hope Creek Generating Station: Unit 1	1,067,000	Public Service Electric and Gas, N.J.	1982
Salem	Hope Creek Generating Station: Unit 2	1,067,000	Public Service Electric and Gas, N.J.	1984
Little Egg Inlet	Atlantic Generating Station: Unit 1	1,150,000	Public Service Electric and Gas, N.J.	1985
Little Egg Inlet	Atlantic Generating Station: Unit 2	1,150,000	Public Service Electric and Gas, N.J.	1987
*	—	1,150,000	Public Service Electric and Gas, N.J.	1990
*		1,150,000	Public Service Electric and Gas, N.J.	1992
NEW YORK				
Indian Point	Indian Point Station: Unit 1	265,000	Consolidated Edison Co.	1962
Indian Point	Indian Point Station: Unit 2	873,000	Consolidated Edison Co.	1973
Indian Point	Indian Point Station: Unit 3	965,000	Power Authority of State of N.Y.	1975
Scriba	Nine Mile Point Nuclear Station: Unit 1	610,000	Niagara Mohawk Power Co.	1969
Scriba	Nine Mile Point Nuclear Station: Unit 2	1,080,000	Niagara Mohawk Power Co.	1982
Ontario	R. E. Ginna Nuclear Power Plant: Unit 1	490,000	Rochester Gas & Electric Co.	1970
Brookhaven	Shoreham Nuclear Power Station	819,000	Long Island Lighting Co.	1978
Scriba	James A. FitzPatrick Nuclear Power Plant	821,000	Power Authority of State of N.Y.	1975
Cementon	Greene County Nuclear Power Plant	1,191,000	Power Authority of State of N.Y.	1983
Jamesport	Jamesport 1	1,150,000	Long Island Lighting Co.	1982
Jamesport	Jamesport 2	1,150,000	Long Island Lighting Co.	1984
Oswego	Sterling Nuclear: Unit 1	1,150,000	Rochester Gas & Electric Co.	1984
*	—	1,200,000	N.Y. State Electric & Gas	1986
*		1,200,000	N.Y. State Electric & Gas	1988
NORTH CAROLINA				
Southport	Brunswick Steam Electric Plant: Unit 1	821,000	Carolina Power and Light Co.	1977
Southport	Brunswick Steam Electric Plant: Unit 2	821,000	Carolina Power and Light Co.	1975
Cowans Ford Dam	Wm. B. McGuire Nuclear Station: Unit 1	1,180,000	Duke Power Co.	1978
Cowans Ford Dam	Wm. B. McGuire Nuclear Station: Unit 2	1,180,000	Duke Power Co.	1979
Bonsal	Shearon Harris Plant: Unit 1	900,000	Carolina Power & Light Co.	1984
Bonsal	Shearon Harris Plant: Unit 2	900,000	Carolina Power & Light Co.	1986
Bonsal	Shearon Harris Plant: Unit 3	900,000	Carolina Power & Light Co.	1988

SITE	PLANT NAME	CAPACITY (Net Kilowatts)	UTILITY	COMMERCIAL OPERATION
Bonsal	Shearon Harris Plant: Unit 4	900,000	Carolina Power & Light Co.	1990
Davie County	Perkins Nuclear Station: Unit 1	1,280,000	Duke Power Co.	1983
Davie County	Perkins Nuclear Station: Unit 2	1,280,000	Duke Power Co.	1985
Davie County	Perkins Nuclear Station: Unit 3	1,280,000	Duke Power Co.	1987
*	—	1,150,000	Carolina Power & Light Co.	—
*	—	1,150,000	Carolina Power & Light Co.	—
*	—	1,150,000	Carolina Power & Light Co.	—
OHIO				
Oak Harbor	Davis-Besse Nuclear Power Station: Unit 1	906,000	Toledo Edison—Cleveland El. Illum. Co.	1976
Oak Harbor	Davis-Besse Nuclear Power Station: Unit 2	906,000	Toledo Edison—Cleveland El. Illum. Co.	1983
Oak Harbor	Davis-Besse Nuclear Power Station: Unit 3	906,000	Toledo Edison—Cleveland El. Illum. Co.	1985
Perry	Perry Nuclear Power Plant: Unit 1	1,205,000	Cleveland Electric Illuminating Co.	1980
Perry	Perry Nuclear Power Plant: Unit 2	1,205,000	Cleveland Electric Illuminating Co.	1982
Moscow	Wm. H. Zimmer Nuclear Power Station: Unit 1	810,000	Cincinnati Gas & Electric Co.	1979
Moscow	Wm. H. Zimmer Nuclear Power Station: Unit 2	1,170,000	Cincinnati Gas & Electric Co.	1984
OKLAHOMA				
Inola	Black Fox Nuclear Station: Unit 1	1,150,000	Public Service of Oklahoma	1983
Inola	Black Fox Nuclear Station: Unit 2	1,150,000	Public Service of Oklahoma	1985
OREGON				
Prescott	Trojan Nuclear Plant: Unit 1	1,130,000	Portland General Electric Co.	1976
Arlington	Pebble Springs Nuclear Plant: Unit 1	1,260,000	Portland General Electric Co.	1983
Arlington	Pebble Springs Nuclear Plant: Unit 2	1,260,000	Portland General Electric Co.	1986
PENNSYLVANIA				
Peach Bottom	Peach Bottom Atomic Power Station: Unit 2	1,065,000	Philadelphia Electric Co.	1974
Peach Bottom	Peach Bottom Atomic Power Station: Unit 3	1,065,000	Philadelphia Electric Co.	1974
Pottstown	Limerick Generating Station: Unit 1	1,065,000	Philadelphia Electric Co.	1981
Pottstown	Limerick Generating Station: Unit 2	1,065,000	Philadelphia Electric Co.	1982
Shippingport	Shippingport Atomic Power Station	90,000	Duquesne Light Co.	1957
Shippingport	Beaver Valley Power Station: Unit 1	852,000	Duquesne Light Co.—Ohio Edison Co.	1975
Shippingport	Beaver Valley Power Station: Unit 2	852,000	Duquesne Light Co.—Ohio Edison Co.	1981

SITE	PLANT NAME	CAPACITY (Net Kilowatts)	UTILITY	COMMERCIAL OPERATION
Goldsboro	Three Mile Island Nuclear Station: Unit 1	819,000	Metropolitan Edison Co.	1974
Goldsboro	Three Mile Island Nuclear Station: Unit 2	906,000	Jersey Central Power & Light Co.	1978
Berwick	Susquehanna Steam Electric Station: Unit 1	1,050,000	Pennsylvania Power and Light	1980
Berwick	Susquehanna Steam Electric Station: Unit 2		Pennsylvania Power and Light	1982
SOUTH CAROLINA				
Hartsville	H. B. Robinson S.E. Plant: Unit 2	700,000	Carolina Power & Light Co.	1971
Seneca	Oconee Nuclear Station: Unit 1	871,000	Duke Power Co.	1973
Seneca	Oconee Nuclear Station: Unit 2	871,000	Duke Power Co.	1974
Seneca	Oconee Nuclear Station: Unit 3	871,000	Duke Power Co.	1974
Broad River	Virgil C. Summer Nuclear Station: Unit 1	900,000	South Carolina Electric & Gas Co.	1979
Lake Wylie	Catawba Nuclear Station: Unit 1	1,153,000	Duke Power Co.	1981
Lake Wylie	Catawba Nuclear Station: Unit 2	1,153,000	Duke Power Co.	1982
Cherokee County	Cherokee Nuclear Station: Unit 1	1,280,000	Duke Power Co.	1984
Cherokee County	Cherokee Nuclear Station: Unit 2	1,280,000	Duke Power Co.	1986
Cherokee County	Cherokee Nuclear Station: Unit 3	1,280,000	Duke Power Co.	1988
TENNESSEE				
Daisy	Sequoyah Nuclear Power Plant: Unit 1	1,148,000	Tennessee Valley Authority	1977
Daisy	Sequoyah Nuclear Power Plant: Unit 2	1,148,000	Tennessee Valley Authority	1977
Spring City	Watts Bar Nuclear Plant: Unit 1	1,177,000	Tennessee Valley Authority	1978
Spring City	Watts Bar Nuclear Plant: Unit 2	1,177,000	Tennessee Valley Authority	1979
Oak Ridge	Clinch River Breeder Reactor Plant	350,000	Project Management Corporation	1982
Hartsville	—	1,233,000	Tennessee Valley Authority	1980
Hartsville	—	1,233,000	Tennessee Valley Authority	1981
Hartsville	—	1,233,000	Tennessee Valley Authority	1981
Hartsville	—	1,233,000	Tennessee Valley Authority	1982
TEXAS				
Glen Rose	Commanche Peak Steam Electric Station: Unit 1	1,150,000	Texas Utilities Services Inc.	1980
Glen Rose	Commanche Peak Steam Electric Station: Unit 2	1,150,000	Texas Utilities Services Inc.	1982
Jasper	Blue Hills: Unit 1	918,000	Gulf States Utilities	1985
Jasper	Blue Hills: Unit 2	918,000	Gulf States Utilities	1987

SITE	PLANT NAME	CAPACITY (Net Kilowatts)	UTILITY	COMMERCIAL OPERATION
Wallis	Allens Creek: Unit 1	1,150,000	Houston Lighting & Power Co.	—
Wallis	Allens Creek: Unit 2	1,150,000	Houston Lighting & Power Co.	—
Matagorda County	South Texas Project	1,250,000	Central Power & Lt.—Houston Lt. & Power	1981
Matagorda County	South Texas Project	1,250,000	Central Power & Lt.—Houston Lt. & Power	1981
VERMONT				
Vernon	Vermont Yankee Generating Station	513,900	Vermont Yankee Nuclear Power Corp.	1972
VIRGINIA				
Gravel Neck	Surry Power Station: Unit 1	788,000	Virginia Electric & Power Company	1972
Gravel Neck	Surry Power Station: Unit 2	788,000	Virginia Electric & Power Company	1973
Mineral	North Anna Power Station: Unit 1	898,000	Virginia Electric & Power Company	1977
Mineral	North Anna Power Station: Unit 2	898,000	Virginia Electric & Power Company	1977
Mineral	North Anna Power Station: Unit 3	907,000	Virginia Electric & Power Company	1980
Mineral	North Anna Power Station: Unit 4	907,000	Virginia Electric & Power Company	1981
Gravel Neck	Surry Power Station: Unit 3	859,000	Virginia Electric & Power Company	1983
Gravel Neck	Surry Power Station: Unit 4	859,000	Virginia Electric & Power Company	1984
WASHINGTON				
Richland	N-Reactor/WPPSS Steam	850,000	Energy Research and Development Admin.	1966
Richland	WPPSS No. 1	1,218,000	Washington Public Power Supply System	1980
Richland	WPPSS No. 2	1,103,000	Washington Public Power Supply System	1978
Satsop	WPPSS No. 3	1,242,000	Washington Public Power Supply System	1981
Richland	WPPSS No. 4	1,218,000	Washington Public Power Supply System	1982
Satsop	WPPSS No. 5	1,242,000	Washington Public Power Supply System	1983
Sedro Woolley	Skagit Nuclear Project: Unit 1	1,277,000	Puget Sound Power & Light	1982
Sedro Woolley	Skagit Nuclear Project: Unit 2	1,277,000	Puget Sound Power & Light	1985
WISCONSIN				
Genoa	Genoa Nuclear Generating Station	50,000	Dairyland Power Cooperative	1971
Two Creeks	Point Beach Nuclear Plant: Unit 1	497,000	Wisconsin Michigan Power Co.	1970
Two Creeks	Point Beach Nuclear Plant: Unit 2	497,000	Wisconsin Michigan Power Co.	1972
Carlton	Kewaunee Nuclear Power Plant: Unit 1	541,000	Wisconsin Public Service Corp.	1974
Ft. Atkinson	Koshkonong Nuclear Plant: Unit 1	900,000	Wisconsin Electric Power Co.	1983
Ft. Atkinson	Koshkonong Nuclear Plant: Unit 2	900,000	Wisconsin Electric Power Co.	1984
Durand	Tyrone Energy Park: Unit 1	1,150,000	Northern States Power Co.	1985

SITE	PLANT NAME	CAPACITY (Net Kilowatts)	UTILITY	COMMERCIAL OPERATION
PUERTO RICO				
Arecibo	North Coast Power Plant	583,000	Puerto Rico Water Resources Authority	—
*Site not selected.				
*	—	1,233,000	Tennessee Valley Authority	1982
*	—	1,300,000	Tennessee Valley Authority	1983
*	—	1,233,000	Tennessee Valley Authority	1983
*	—	1,300,000	Tennessee Valley Authority	1984
*	—	1,150,000	New England Electric System	1983
*	—	1,150,000	New England Electric System	1985

INDEX

Index

ABOUT THE AUTHOR

McKinley C. Olson is a Chicago-based editor-writer-photographer and a former editor and editorial writer for the York (Pa.) *Gazette and Daily*. His first book was *J. W. Gitt's Sweet Land of Liberty*. Olson has written about nuclear power for a number of publications, including *The Nation* and *The Progressive*.

Bring out the books that bring in the issues.

Bantam Book Catalog

It lists over a thousand money-saving best-sellers originally priced from $3.75 to $15.00 —bestsellers that are yours now for as little as 60¢ to $2.95!

The catalog gives you a great opportunity to build your own private library at huge savings!

So don't delay any longer—send us your name and address and 25¢ (to help defray postage and handling costs).